PRACTICAL STABLE
MANAGEMENT

PRACTICAL STABLE MANAGEMENT

Christine E. Hughes and Robert Oliver
Diagrams by Dianne Breeze

PELHAM BOOKS
London

First published in Great Britain by
Pelham Books Ltd
27 Wrights Lane
London W8 5TZ
1987

British Library Cataloguing in Publication Data

Hughes, Christine
    Practical stable management
    1. Horses
    1. Title   2. Oliver, Robert, *1940–*
    636.1'083   SF285.3
    ISBN 0-7207-1759-0

Typeset in Ehrhardt by Cambridge Photosetting
Services
Printed and bound in Great Britain
by Butler and Tanner Ltd., Frome, Somerset

CONTENTS

AUTHORS' NOTE

Throughout this book measurements are expressed in both metric and imperial units except in cases where standard practice and/or conventions in the horse world dictated otherwise.

"After man; the most eminent creature is the horse;
The best employment is that of rearing it;
The most meritorious of domestic actions is that of feeding it;
And the most delightful posture is that of sitting on its back."

ACKNOWLEDGMENTS

The authors would like to thank the following people for generously giving of their help and time towards compiling this book: Alexandra Oliver, David S. Hughes and John M. Killingbeck, B.Sc., B.V.M.S., M.R.C.V.S. We are particularly indebted to Dianne Breeze for her wonderful illustrations which have brought life to these pages. The indefatigable Lesley Gowers whose skills as an editor have transformed a garbled manuscript into an orderly text and John Beaton who cleverly stage-managed the final production. Not forgetting the publishers themselves who have shown their confidence in us yet again.

The authors and publishers are grateful to the following for permission to reproduce illustrations in the book: Christine Hughes photographs on pages 42, 76, 83, 93, 94, 95, 96, 105, 108, 120, 124, 125 (top left), 171; Bob Langrish photographs on pages 78, 80, 82, 107, 125 (top and bottom right); Francesca Spano photographs on pages 66, 68, 172, 177, 189, 190, 191. The diagrams on pages 23 and 26 are based on those in *Pony* Magazine drawn by Christine Bousfield; the diagram on page 135 is reproduced from *Equine Nutrition* by A. C. Leighton Hardman (Pelham Books) and the diagrams on pages 100 and 101 are based on photographs by Donald Tuke in *Bit by Bit* by Diana R. Tuke (J. A. Allen Ltd.)

PREFACE

It will be a source of pride and gratification to the authors that they have been enabled to promote the interests of those who keep horses, as well as to ameliorate the treatment of the noble animal which forms the subject of this book.

Christine E. Hughes and Robert Oliver

INTRODUCTION

A common understanding of the horse, based on a language which is universally and socially accepted, is the foundation of our knowledge of him. As we go about our day-to-day business in the stable this language will become commonplace to us and our vocabulary will extend as new words and terms are introduced. The basis, however, will always be the deeply traditional language of the horse world which is part of our equestrian heritage. As in every sport one is soon identified as a serious participant if one can speak the language; conversely anyone pretending to be knowledgable is soon found out by his ignorance of certain terms and expressions. It is essential then that anyone involved with horses understands horse talk and horse sense if he is to enjoy horses and not be taken for a ride.

The book will begin by offering the reader a broad historical background to the horse and his role as a domesticated animal for our pleasure and utility. This is followed by an explanation of horse management and general equestrian knowledge required by the horseman in preparation for the more specialised subjects to follow. The duties and virtues of a groom are examined in some detail but horsemanship as such is a more inherent instinct. Correct and appropriate guidance can develop and enhance a basic natural ability and sensitivity. It cannot, however, be taught to any individual who lacks a fundamental and essential sympathy and understanding of the horse as an animal.

Instinct plays a large part in horsemanship and much of a horseman's communication with the horse is achieved through a kind of sixth sense – in other words, an inborn anticipation of the horse's reactions to a given situation. Not only must the horseman understand how a horse may react but his natural intuition will automatically prompt an appropriate response to pacify a frightened animal or reprimand his misbehaviour. Any reaction to a situation on the part of the horse will be spontaneous and require an immediate answer from the handler if the moment is not to be lost in indecision. Every situation is different and must be dealt with individually, as indeed must each horse. There is very little room for theory in educating the novice horseman to deal appropriately and effectively with every circumstance, only experience built on the qualities mentioned above will establish a sound and practical knowledge.

INTRODUCING THE HORSE

The equine species

There are six members of the equine species as we know it today, which all come under the collective name of *Equus*. They are:

The onager (*Equus hemionus onager*) – a type of wild ass found in Central Asia from which local strains have been bred.

The ass – either wild or domesticated. The wild ass is used as a pack animal and for riding, mainly in Asia and Africa. The domesticated breed of the ass family is the donkey which has established itself as a popular pet being used more for pleasure than for work in modern times.

The zebra – there are three different varieties: the Mountain Zebra, the Quagga and Grevy's Zebra, all of which inhabit the African continent and can only be found in captivity elsewhere.

The horse – in all its various breeds.

Of these the horse has played the most significant part in the history of man. In contrast to his cousin the Zebra he has many qualities which make him attractive to us. His intelligence is the chief factor for our intimate relationship with him. He has learnt to trust man and to work closely with him. His sensitive nature and alertness of mind make him easily educated and thus suitable for exploitation. In fact his superior mentality makes him more trainable than any other animal except the dog, while physically his athleticism and agility are qualities which other large animals cannot match and will not share with man. Partnerships between man and horse have been part of our heritage for thousands of years – something of which we can be justly proud. Out of our relationship with him is born a mutual respect and affection which non-equestrians may find difficult to comprehend.

In his wild state the horse has adapted to running on the hard soils of the plains eating harsh grasses. He is, however, not necessarily a steppe animal as can be seen by his presence in forest and tundra areas. By nature he is a gregarious animal used to living in herds of six to twenty animals, each group with its own lead stallion. His speed and agility enable him to escape from predators with a quickness of flight, and he is ever vigilant. His sensitive eye, whose field of vision can sweep in a wide semi-circle without the need for him to move his head, ensures that he can graze whilst being fully aware of his surroundings. The shape of the horse's hoof contributes to his survival in the snow because it enables him to paw away at the icy covering to expose the grass beneath. Conversely it is less efficient in soft, sandy or swampy soils. Likewise he is at a disadvantage in mountainous terrain because he cannot rely on the same degree of surefootedness as the sheep and goat with their cloven hooves.

The stallion of a herd is polygamous and will not tolerate competition, even from the colts within the herd. Once the young males reach maturity the lead stallion will herd them off to ensure that he has his 'wives' totally to himself. Whilst protecting his own mares he is constantly seeking to attract females from other herds and retain as many as he can, including his own daughters. In guarding his wives it is more accurate to say that he drives rather than leads them, which characterises his dominant nature. Amongst themselves

11

horses have a natural instinct to establish this dominance and they have a kicking and biting order just as poultry have a pecking order. Within their groups grooming, in the form of nibbling, is an important part of socialising. When horses are turned out together in paddocks they can often be seen nibbling each others' withers and backs.

The history of the horse

The earliest estimated date of the domestication of the horse is 3,000 BC. The dog, in comparison, dates back to 12,000 BC. It is thought that cattle, discovered some 3,000 years earlier than the horse, led the way in serving as a model. The reasons for his domestication are complex but essentially he was recognised as a source of meat. Economy and ideology also played a part for the horse was able to make efficient use of pastures all the year round. As well as for human consumption horse flesh was used for sacrificial purposes and as part of burial rites. It follows then that man soon realised the need to contain the horse in an area from which left to himself he could not roam and so corral-type yards were constructed for his enclosure. It was necessary to tame the horse in order to have some degree of control and to use him as a decoy to attract other horses. Castration was carried out for these reasons. Riding was soon discovered to be essential as a means of herding wild horses and to effect a measure of control, bits, initially made of bone, were brought into use. At this time there were no saddles but it is believed that a cloth or pad served as a means of protection and security for the rider. It is thought that the horse was harnessed to a cart of some sort too, which suggests an early understanding on the part of man as to the adaptability of the horse.

The size of some breeds has changed over the years resulting in the modern animal being somewhat larger than his ancestors. It is evident from the size of early bridles, bits and halters that his head has altered in shape too. His forehead is broader while the volume of the braincase has actually decreased. The facial part has shortened and the muzzle has narrowed down, making the teeth a smaller size. We are told that even from the earliest times the horse was stabled and consequently fodder had to be collected. From such records we can see that stable management is nothing new although it has come a long way since then.

One of the most important roles of the horse through the centuries has been his participation in war. He was used as a vehicle by early nomadic herdsmen before his introduction to the chariot and to the cavalry. As a consequence of his changing role a larger animal was bred to carry greater weights because the first horses were too small and light for these tasks. As the horse became larger and heavier he was also put to work in many ways, such as in the fields, pulling carts etc. As a war horse his involvement is well recorded but it bears repeating that his services to man during combat had a significant effect on man's history. Socially, too, his ownership was soon established, with the better horses being kept by the aristocrats and the lesser animals carrying out more mundane duties for the peasants.

It is interesting to note, especially in these modern times of everchanging trends, that some things have not altered for thousands of years, in particular some feedstuffs. Chariot horses, for example, were fed on barley, wheat, alfalfa and chopped straw. During medieval times bits became widely used with the distinction between curbs and snaffles soon identified. One reason for curbs becoming popular was so that the knight, who held his reins high and was mounted on a saddle with a high pommel and cantle, could control his horse with a minimum of effort. His saddle had to be so designed to give him greater security – once he fell he would be defenceless. Curb bits came to signify nobility since the lower classes rode in snaffles. (Perhaps this custom was carried into the hunting field because traditionally a curb or double bridle is worn by foxhunters today.) The knight also demanded a stronger horse to carry the great weight of his armour, which could add up to

30 stones mounted, and the lighter saddle horses were not up to the weight or the demands of charging at full gallop. It was after the thirteenth century that heavier horses really became widespread although Richard the Lionheart apparently rode moderate-sized horses of Turkish and Cypriot descent. By the eighteenth century the larger draught horses were being employed for pulling carriages and coaches. The armoured knight was made redundant by the invention of gunpowder but his sport of quintain (jousting) lived on and still today we can enjoy the pageantry of medieval times exhibited in the form of tournaments.

Equestrian sport as such originated in Rome, Greece and central Asia but our medieval ancestors also practised mounted games. Bullfighting was another early sport in which the Spanish employed horses and training took between four and six years before the horse would be ready to enter the arena.

The modern horse is mainly kept for sporting and leisure purposes as his role as a working vehicle on the farm has been largely superseded by the tractor. He is, however, still used on the land in primitive countries and where farming is carried out on a smaller scale.

Outside the sporting arena the main employers of the horse are the police and the cavalry, but horses can also be found pulling drays, working the land, performing in circuses and serving on cattle stations. The main sporting pursuits are dressage, show jumping, showing, driving, eventing, triathlon, polo, long-distance riding, gymkhanas, Pony Club activities, vaulting, jousting, tent pegging and, of course, hunting the fox, deer, otter, stag and drag.

Behavioural patterns of the domestic horse in enclosed surroundings

Given that we are taking a wild animal and confining him in an unnatural environment purely for our own indulgence, we should expect to see changes in his behaviour from that which is normal to him in his state of freedom. What we are so often guilty of is ignoring signs of stress in the horse, however minimal, for the sake of our own selfish convenience. However generous a horse's temperament we should never forget that every minute spent in the stable causes a state of physical stagnation in an active animal and a sense of isolation which goes against his gregarious nature. Any attempt by the horse to actively while away the long and tedious hours spent indoors is regarded as a vice. Man is constantly and often unnecessarily suppressing the horse's expressions of feelings, and in horses of a less generous nature this can provoke further frustrations and habits which man will reprimand, sometimes violently.

The majority of stable vices are caused by boredom and frustration born out of compulsory idleness. It is therefore entirely the responsibility of man to ensure that everything possible is done to alleviate this state of discontent. Some experienced equestrians would claim that solitary permanent confinement to the four walls of a stable, broken only by a period of exercise, is nothing less than cruelty. Why should we expect a horse to behave mechanically? Witness the effect of confinement on man. The horse may not be of equal intelligence but any prolonged inactivity will cause staleness. Thus it is advisable that whenever possible the horse should be allowed to roam freely at pasture to break up the monotony of the day once he has worked. Without the facilities to turn horses out the groom should be prepared to walk out his charges at intervals to stimulate the animals' circulation and give them the chance to have a bite of grass because, after all, that is the most natural occupation for the horse.

When a horse is first brought into enclosed surroundings his bewilderment may cause problems. However with correct and sympathetic handling the difficulties are often short-lived and the horse soon learns to adjust. If, on the other hand, the handler mistreats the horse the situation can develop into something serious. An experienced horseman will be aware of the consequences of mishandling a difficult horse but the novice, through inexperience, may induce a permanent behavioural problem, especially in a young horse.

Breeds

There are in the region of two hundred breeds in the world today, some of which have subbreeds. Breeds can be divided into four categories depending on the type of animal. These categories are known as:

Hot Bloods (HB) – lighter breeds.

Cold Bloods (CB) – heavier breeds such as draught animals.

Warm Blood (WB) and Ponies (P) – a combination of the previous two.

In the United Kingdom there are eleven native breeds or types:

Connemara	New Forest
Dales	Shetland
Dartmoor	Welsh Cob
Exmoor	Welsh Mountain
Fell	Welsh Pony
Highland	

Some of these still run wild and many have been cross-bred with other breeds and types to produce a combination of qualities.

The following is a list of breeds which are most popular in the British Isles today:

Andalusian – WB
Anglo-Arab – HB
Appaloosa – HB
Argentina – HB
Australian – WB
British Warm-Blood – WB
Cleveland Bay – CB
Clydesdale – CB
Connemara – P
Dales – P
Danish Warm-Blood – WB
Dartmoor – P
Exmoor – P
Fell – P
Hackney Horse and Hackney Pony – WB

Hanoverian – WB
Highland – P
Irish Draught – CB
Irish Horse – WB
Lippizaner – WB
Morgan – WB
New Forest – P
Percheron – CB
Shetland – P
Shire – CB
Suffolk – CB
Thoroughbred: British, French, Italian, North American – HB
Welsh: Cob – CB
Pony: Cob type – CB
      Mountain – CB
Westphalian – WB
Yorkshire Coach Horse – CB

Of all the breeds known to us perhaps the Shetland, Shire and Thoroughbred are the most familiar representatives of their types.

As a result of many years of cross-breeding we now have a number of types of horses and ponies suitable for many different uses and disciplines. For example, the Irish Draught × Thoroughbred is proving a successful marriage for producing competition horses which have the courage and athleticism necessary for the modern sports.

In recent years, breed societies have introduced stringent regulations to ensure that animals which have serious conformation faults or any history of disease or unsoundness which could be inherited are not bred from. Thus each society hopes to preserve the qualities and characteristics of its chosen breed.

Conformation

The following points are general guidelines only and do not take into account breed characteristics.

Head – to begin with, the horse's head should look as if it 'belongs' to its owner, i.e. it should be in proportion with the rest of the body. Many people give priority to the head when selecting a horse: even if the horse proves to be

POINTS OF THE HORSE.

1 POLL
2 CHEEK
3 MUZZLE
4 CHIN
5 JOWL
6 CREST
7 NECK
8 JUGULAR GROOVE
9 WITHERS
10 SHOULDER
11 POINT OF SHOULDER
12 ARM
13 BREAST
14 POINT OF THE ELBOW
15 BRISKET

16 FOREARM
17 KNEE
18 FORE CANNON
19 FETLOCK
20 CORONET
21 CHESTNUT
22 ERGOT
23 PASTERN
24 HOOF
25 CHEST
26 BACK
27 BELLY
28 FLANK
29 SHEATH
30 LOINS

31 CROUP
32 DOCK
33 HIP JOINT
34 HIND QUARTERS
35 THIGH
36 STIFLE
37 BUTTOCK
38 GASKIN
39 HAMSTRING
40 POINT OF THE HOCK
41 HOCK JOINT
42 CHESTNUT
43 HIND CANNON
44 FETLOCK
45 BULB OF HEEL

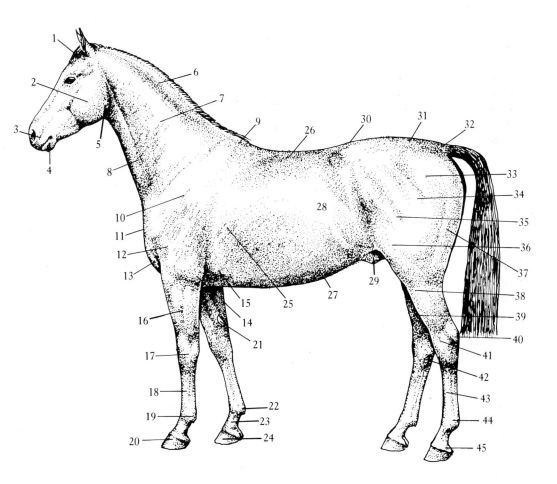

a good performer his attraction is sometimes lost if he has a large, plain, ugly head. A horse's character and disposition are often judged by his head and it often follows that a horse with a common head carries it badly. Qualities to look for are a nice lean head with a broad, flat forehead, no large bumps between the eyes and large open nostrils.

Eyes – a bold eye is a sign of intelligence and courage while a small 'piggy' eye is an indication of meanness and wickedness. Horses which show the white of their eye are regarded as being bad tempered.

Ears – undoubtedly one of the most attractive features of a horse's head are ears which are constantly pricked forwards and upright, signifying interest and intelligence. Long, loppy ears are often found on Thoroughbreds and while many would not regard them as attractive they usually indicate that the horse is a good performer, although the connection is somewhat unclear.

Neck – the way in which the head is set on the neck and how the horse carries his neck can affect how the horse goes. For example, the horse who carries his head low is often an uncomfortable ride, particularly if he has an upright shoulder. A very long, low neck is unattractive and it invariably follows that the horse will go on his forehand. A 'ewe' or upside-down neck is due either to a conformation fault or to bad schooling. If a horse has been allowed to go with his head held high he will develop the muscles on the underside of his neck. In such cases retraining can gradually correct the problem. A short, thick 'cobby' type neck often predisposes respiratory problems, i.e. the horse will make a noise in his wind. A good neck should give the appearance of a long front which will offer plenty in front of the rider and a nice long rein. It should be well shaped and run into a nicely sloping shoulder.

Shoulder – to be a good ride the shoulder should not be narrow otherwise it will be sharp and uncomfortable, rather like riding a gate. On the other hand coarse, very wide shoulders on a common horse are not pleasant to ride on either. If the horse's neck and shoulders are not right the saddle will not sit in the proper place and will move forwards.

Withers – high withers, which are often found on Thoroughbred horses, can at times make saddle fitting difficult.

Back – the back should be level, short, strong and muscular. It should not be dipped or swayed in any way because although this may be comfortable to ride on it is a weakness. A roach back is both uncomfortable and a fault. To hold his condition during work the horse must have well-sprung ribs, otherwise he will look weak and poor. Horses with flat or greyhound-like sides should be avoided as should 'herring-gutted' animals because these run up light when they are in work and need a breastplate to keep the saddle in place. No amount of food or schooling will correct these faults. A deep girth always goes in the horse's favour as it indicates plenty of heart room. The bump which some horses have behind their loins – known as 'jumper's bump' – is ugly and serves no purpose. A goose rump is one which slopes away to a low-set-on tail, and is undesirable. The hindquarters should be level and strong and complemented by a well-shaped hind leg, for it is this part of the horse which serves as the 'engine'.

Loins – muscular loins are naturally better than those which are hollow and therefore weak and often go together with a long back. Correct schooling can help to develop weak loins but if the horse's basic conformation is inadequate nothing will rectify it.

Tail – a nicely set-on tail which is carried well will look attractive. It is a characteristic of the Arab horse to carry the tail quite high indeed; some other foreign breeds have tails which are set on high and carried straight out.

Fore legs – one should look for big, flat knees, not round or small. Also desirable are short cannon bones, flat fetlocks, and short pasterns with a normal slope of 45° which will give a comfortable ride because they act as shock absorbers. Long, sloping or upright pasterns are a weakness and often predispose lameness of the fore limbs.

A horse which has a forward bend of the knee joint is described as being 'over at the

THE SKELETON OF THE HORSE

1 TEMPORAL FOSSA
2 FRONTAL BONE
3 ORBIT
4 MAXILLARY BONE
5 NASAL BONE
6 INCISIVE BONE
(PREMAXILLARY)
7 MANDIBLE
8 ATLAS (FIRST CERVICAL
VERTEBRA)
9 AXIS (SECOND CERVICAL
VERTEBRA)
10 CERVICAL VERTEBRA
(TOTAL SEVEN)
11 SCAPULAR SPINE
12 SCAPULAR
13 SCAPULAR CARTILAGE
14 SHOULDER JOINT
15 HUMERAL TUBEROSITY
LATERAL

16 HUMERUS
17 STERNUM
18 OLECRANON
19 ELBOW JOINT
20 ULNA
21 RADIUS
22 CARPUS
23 METACARPUS
24 FETLOCK JOINT
25 COFFIN JOINT
26 SMALL METACARPAL
BONE
27 PROXIMAL SESAMOID
BONE
28 PROXIMAL PHALANX
(FIRST PHALANX)
29 DISTAL PHALANX (THIRD
PHALANX)
30 RIBS (TOTAL EIGHTEEN)
31 COSTAL CARTILAGES
32 THORACIL VERTEBRAE
(TOTAL EIGHTEEN)
33 LUMBAR VERTEBRAE
(TOTAL SIX)
34 COSTAL ARCH
35 TUBER COXAE

36 TUBER SACRALE
37 SACRAL VERTEBRAE
(TOTAL FIVE) FUSED
TOGETHER
38 ILIUM
39 PUBIS
40 HIP JOINT
41 FEMUR (GREATER
TROCHANTER)
42 TUBER ISCHII
43 ISCHIUM
44 FEMUR (THIRD
TROCHANTER)
45 STIFLE JOINT
46 FIBULA
47 CALCANEUS (FIBULAR
TARSAL BONE)
48 TARSUS
49 SECOND PHALANX
(MIDDLE)
50 TIBIA
51 SMALL METATARSAL
BONE
52 METATARSUS
53 EASTERN JOINT
54 PATELLA

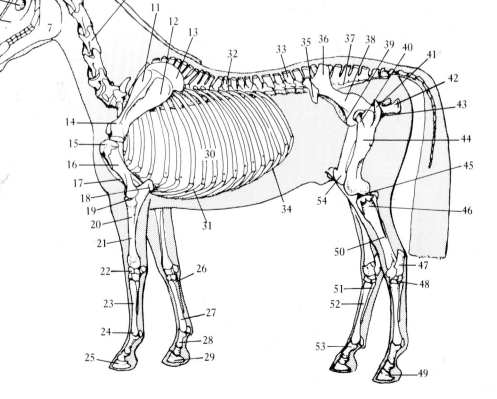

MUSCLES OF THE HORSE.

1 SPLENIUS
2 SERRATUS MAGNUS
3 TRAPEZIUS
4 LATISSIMUS DORSI
5 GLUTEUS ESTERNUS
6 FASCIA LATA
7 BICEPS ABDUCTOR
 FEMORIS
8 SEMI-TENDINOSUS
9 SEMI-MEMBRANOSUS
10 RECTUS ABDOMINUS

11 OBLIQUUS ABDOMINUS
 EXTERNUS
12 INTERCOSTAL MUSCLES
13 SERRATUS MAGNUS
14 TRICEPS EXTENSOR
 BRACHII
15 TERUS EXTERNUS (LONG
 ABDUCTOR OF THE ARM)
16 POSTEA SPINATUS
17 ANTEA SPINATUS
18 PECTORALIS PARVUS
19 STERNO MAXILLARIS
20 MASTOIDO HUMERALUS

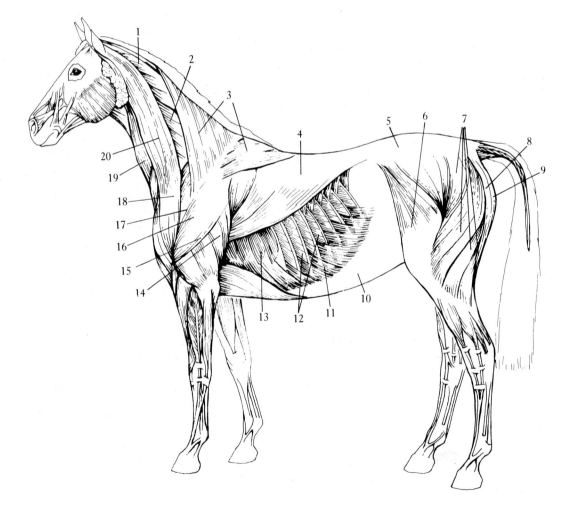

knee'. There are varying degrees of bend on the joint and in less extreme cases strain on the tendon is minimal. It is not an unsoundness and opinion suggests that a horse which is over at the knee will not break down. It is often found in Thoroughbred horses.

Being 'back at the knee' is the opposite of the above and is considered a conformation fault making the animal unsuitable for showing. In contrast to over at the knee this fault is more often found in common-bred, cob-type horses.

Tendons – these should be hard and sinewy and not in any way soft or puffy.

Feet – the old adage 'no foot, no horse', should always be borne in mind when selecting a horse. The horse's feet should correspond to the size of the animal and should be neither narrow and boxy nor flat and wide. So many problems are connected with the feet that one cannot place too much emphasis on them.

Hind legs – the hind leg of a horse should, when viewed from the side, be directly below him, i.e. follow a line from the point of buttock to point of hock and back of fetlock joint. It should also have a well-developed second thigh. Two main faults of the hind leg are the sickle hock – one which stands underneath the animal; and the cow hock – one where the hocks are closer together than the feet.

Hocks – there should be no swelling around this joint and it should appear clean and fibrous. Curbs, spavins and thoroughpins are some of the chief problems relating to this complicated joint, some of which are due solely to bad conformation.

Having gained a general picture of a horse and established that it has no major conformation faults and is a good basic shape, you should then evaluate the horse bit by bit starting with the feet. A pretty horse with an attractive head should not blind a good horseman. Each horse is only as good as his weakest point and must be sound if he is to perform. Never ignore a fault, especially in the limbs, because a problem will inevitably manifest itself sooner or later when the horse is put to work.

Movement and action

True, straight movement of the horse is always looked for when assessing him for both soundness and correctness. The animal that is level and straight should wear better and last longer on its legs and joints than one that has irregularities such as dishing and crooked legs or bad feet. When viewed from behind the horse is described as having either a wide or narrow base and the ideal is the horse in between the two with no signs of brushing or plaiting of the legs. Horses which turn their feet out as in plaiting are less likely to injure themselves than those who dish, i.e. turn them in.

Different movements are required for the various jobs which the horse is to carry out, i.e. the show horse is expected to have long, low action whereas the show jumper needs a shorter stride with higher knee action. In the racehorse action comes in all shapes and sizes; often the trot is disappointing but they usually walk well – in fact Thoroughbred yearlings are only ever expected to walk at sales.

The recognised way of examining the horse's movement or action is first of all to stand in front of the horse as he walks towards you, then directly behind him as he walks away, and finally at right angles to him as he passes. At no time when the horse is moving should you be able to see the sole of his foot which often happens when the heel is put down first. The horse should move easily in all his paces covering the ground with even strides. He must not trail his hocks behind him, lower his head or hit the ground heavily.

Sex

In general conversation the word 'horse' broadly describes the animal of either sex but to be more precise the following terms are used.

Filly – a female horse as a yearling, two-year-old and three-year-old.

Mare – a female from four years of age or when in foal, whichever comes first.

Colt – an uncastrated male under three years old.

19

Gelding – a castrated male of any age.

Stallion or **Entire** – an uncastrated male of four years of age and over.

Rig – a male horse from which only one testicle has been removed. It is sometimes found whilst castrating a colt that only one testicle has dropped down into the scrotum. More often than not this is rectified in time by nature and removal is successfully carried out at a later date.

Colours

The horse comes in many different colours. It is important to learn to identify these and to be able to distinguish the differences between when a horse is in his summer coat and when he has been clipped, because the tone of some colours may change slightly. There is usually an alteration in the coat colour of a foal during its first year and this applies more to some colours than to others. Grey animals are known particularly for their change of colour over the years; as they get older the whiter they become. The horse's muzzle is referred to when identifying his true colour. Personal colour preference plays a large part in the selection of a horse but breeders, of course, have no choice. As with the sex of the horse, buyers can be quite emphatic in their choice, basing their preference on experience, but with some colours and markings, tradition and hearsay also play a part. It is said, for example, that four white socks are an indication that the horse will not be a good one. A strong colour is generally considered more attractive and desirable; an old saying suggests 'washy in colour, washy in constitution'. But, like everything else, there are always exceptions.

The bay (b) comes in three different shades – light (yellowish/reddish brown), bright (dark chestnutty brown) and dark (rich, dark brown) and has a black mane and tail and usually black on the limbs. It is perhaps the most popular colour of all.

The brown (br) horse is a darker shade of the dark bay but not as dark as the black horse.

The chestnut (ch) is again found in different shades – dark, liver, chestnut and light chest-nut which can be quite pale. The light types often have a flaxen mane and tail while the darker chestnuts usually have a chestnut mane and tail.

The black (bl) although less common than the former colours is especially attractive if the horse has a bloom to his coat. It is easily identified and should have dark points as well as a black muzzle.

The grey (gr) can represent anything from a pure white to a steel or iron colour and can be dappled in varying degrees. The obvious disadvantages of a grey are in grooming and in the skin pigmentation being relatively more prone to skin diseases. The grey's coat invariably becomes lighter in colour as he grows older, sometimes to the extent of becoming completely white.

The roan (r) colours are strawberry or chestnut roan, which is a chestnut colour mixed with white hairs; bay or red roan, which has the body colour of a bay with white hairs; and roan or blue roan which is a brown or black colour with white hairs. All these colours usually have black points, i.e. from the knee and hock down.

The above colours are internationally accepted but the following are less common and do not enjoy the same recognition.

The dun can vary from a mousy colour to a more golden tone and has black points; sometimes there is a black line down the back.

The piebald is a black and white animal.

The skewbald is generally brown and white although technically it can be any other colour except black. With both piebald and skewbald there is no uniformity in the colouring.

The albino has white hair and the skin is pink.

The palomino comes in various shades of gold and has a flaxen mane and tail.

The cream is between the albino and palomino colour, with a pink skin.

Markings

There are a number of identifying marks which can be found on the horse and which have to be recorded on veterinary certificates, passports and when describing the animal for registration or selling.

On the head:

Blaze – a wide covering of white hair on the forehead, sometimes running down into the muzzle.

White-faced or **bald-face** – terms used when the whole face is white.

Star – a roughly round area of white hair of any size in the centre of the forehead.

White hairs (on the forehead) – these should be noted as such if there are only a few and not enough to constitute a star.

Race or **stripe** – a narrow strip of white running down the face in various degrees.

Snip – a white mark between the nostrils.

White muzzle – nothing less than the whole area covering both lips and nostrils.

Underlip and **upperlip** can have white skin and should be noted as such.

On the body:

Stripe, list and **ray** – names given to the dark lines occasionally found along the horse's back but more often on donkeys and mules.

Saddle and **girth marks** – areas of white hair caused by pressure from saddlery. They are very common in older horses and in those subjected to ill-fitting tack.

Flea-bitten – the description given to a coat which has small tufts of white hair on the body. In the case of grey horses these tufts can be black or brown, hence the term 'Flea-bitten grey'.

On the legs:

Each leg (or legs) which is marked should be individually noted in the following ways as appropriate:

White feet.

White heels.

White coronets – if black or chestnut marks appear on white coronets their position should be described.

White pasterns.

White fetlocks.

White socks – white hair from the coronet to just above the fetlock.

White stocking – white hair extending from the coronet up the leg to the knee.

White leg is so called when it runs up high into the forearm.

Black or **ermine spots** – marks found on white areas.

Black points – the term used when the lower part of the leg and hoof is black.

Horses without any white marks on their body are called whole coloured. The position of white marks is always recorded.

Other identifiable marks are:

Whorls – small areas where the hair turns in the opposite direction to the rest of the coat which should be detailed for their position and, where necessary, the length. They are usually found on the forehead, on the crest of the neck, either side, and from the point of the hip down the flank. Most horses have a hair-whorl somewhere.

Prophet's thumb marks – small indentations in the muscle, sometimes found in the shoulder and neck area.

Wall eye – this term describes an iris which is lacking in pigment.

White of the eye – when the cornea of the eye is short of pigment the horse is referred to as showing the white of his eye. It is often said to be indicative of a bad temperament and for this reason some people will avoid buying an animal which has one.

Flesh marks – areas of hairless skin without any pigmentation.

Scars – all permanent scars should be recorded with their exact location and whether they are hairless or not.

Brands – these denote the breed and/or country of origin and are generally found on the shoulder or the quarters but sometimes on the neck or under the saddle.

Freeze-marking – a form of branding which is effected by freezing the skin and is normally located under the saddle. The horse is given a number which is recorded on a national register and made available to the police in the event of theft.

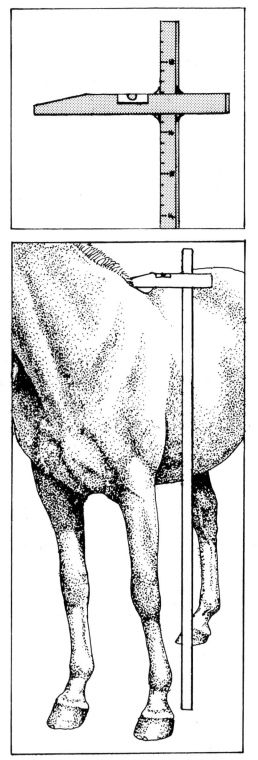

Height

Horses and ponies are measured at the highest part of their withers when they are standing on level ground on all four legs. The thickness of the shoe, e.g. half an inch, should be deducted from the height measured. For showing purposes all horses and ponies must be measured without their shoes by a veterinary surgeon and officials.

Horses and ponies are measured in hands; a hand is 4 inches. A pony does not exceed 14.2 hh (58 ins or 148 cm). Any animal over that is classified as a horse. An exception to this rule is in the decription of polo ponies which although so called, are often of horse size.

A measuring stick is marked in hands, inches and eighths of an inch and has a wooden T-piece which should be fitted with a spirit level. The T-piece slides up and down the shaft of the stick and when lowered onto the withers will give an accurate measurement.

Age and dentition

The horse's age is dated from 1 May, or in the case of Thoroughbreds 1 January, regardless of the actual date of birth. The term 'foal' describes an animal of either sex up to the age of one year, after which he is referred to as a yearling until he is two. Subsequently he is known as a two-year-old, three-year-old, etc. After the age of twelve he is regarded as 'aged'. In order to be able to 'read' a horse's mouth and determine the animal's age it is essential to be familar with the form and arrangement of the teeth.

A horse's age is indicated with accuracy up to eight years of age by the appearance of the incisor teeth. After eight years the shape of these teeth varies so one can only estimate the age to within a year or two. It is said that after the age of twelve it is impossible to tell the age of a horse by looking at his teeth.

The horse has two complete sets of teeth in his life, beginning with temporary or milk

TOP LEFT *Measuring stick with spirit level*
LEFT *How to measure a horse*

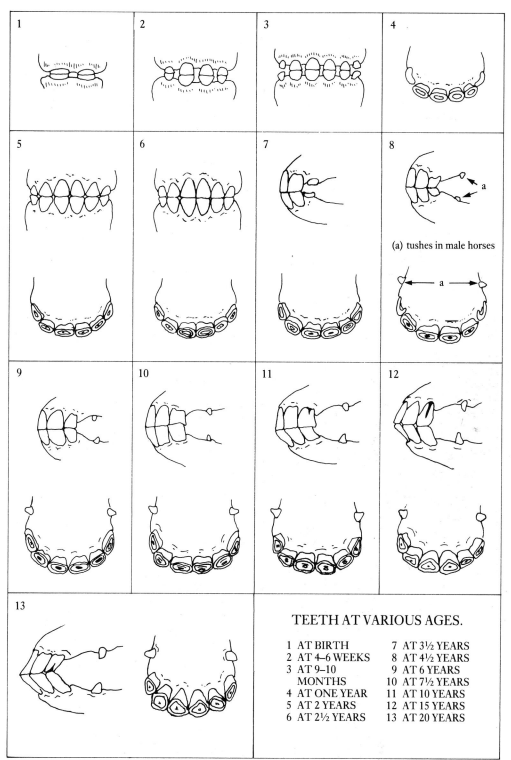

(a) tushes in male horses

TEETH AT VARIOUS AGES.

1 AT BIRTH	7 AT 3½ YEARS
2 AT 4–6 WEEKS	8 AT 4½ YEARS
3 AT 9–10 MONTHS	9 AT 6 YEARS
	10 AT 7½ YEARS
4 AT ONE YEAR	11 AT 10 YEARS
5 AT 2 YEARS	12 AT 15 YEARS
6 AT 2½ YEARS	13 AT 20 YEARS

23

THE HORSE

Molars.

AGES.	LOWER JAW—*Incisors.*			UPPER JAW—*Incisors.*			TEMPORARY.		PERMANENT.			
	CENTRAL.	MIDDLE.	CORNER.	CENTRAL.	MIDDLE.	CORNER.	1.	2.	3.	4.	5.	6.
At Birth,	Temporary 2 very wide.	………	………	………	………	………	At birth,	At birth, and very large.				
2 to 3 Wks.	………	Temporary 2 appear, very wide.	………	………	………	………	………	………	Appears			
6 Months,	………	………	Temporary 2 appear.	………	………	………						
7 to 8 Months,	………	………	Level.	………	………	………						
11 Months,	………	………	………	………	………	………	………	………	………	Appears		
12 Months,	Cups leave.	………	………	………	………	………						
18 Months	Cups leave.	………	………	………	………	………	………	………	………	………	Appears	
2 Years,	………	Cups leave.	Cups leave.	………	………	………						
2½ Years,	Permanent appear.	………	………	………	………	………	Perma-nent appear.	………	………			
3 Years,	Level.	………	………	………	………	………	………	Perma-nent appear.	………	………		Appears
3½ Years,	………	Permanent appear.	………	The Permanent Incisors in Upper Jaw, as a rule, are developed in advance of those in the Lower.								
4 Years,	………	Level.	………									
4½ Years,	………	………	Permanent appear.						Perma-nent appear.			

Years								
5 Years,				Level.	Much wider than long deep notch in inside wall.			Level.
6 Years,	Cups small.			Inside and outside wall level.	Little wider than long.		Small.	Level.
7 Years,	Cups gone.	Cups small.		Inside wall shows wear.	Square or longer than wide.			
8 Years,	Cups gone.	Cups small.	Cups small.		Notch nearly worn out.	Irregularly formed groove appears.		
9 Years,	Cups gone.	Cups gone.	Cups small.	Cups gone.	Inside and outside wall level.	Irregularly formed groove appears.		
10 Years,	Cups gone.	Cups gone.	Cups gone.	Cups gone.	A regularly formed groove appears.			
13 Years,	Cups gone.	Cups gone.	Cups gone.	Cups gone.	About quarter of way down.			
15½ to 16 Years,	Cups gone.	Cups gone.	Cups gone.	Cups gone.	Half-way down.			
About 18 Years,	Cups gone.	Cups gone.	Cups gone.	Cups gone.	Three-quarters of way down.			
21 Years,	Cups gone.	Cups gone.	Cups gone.	Cups gone.	Reaches the bottom.			
22 Years,	Cups gone.	Cups gone.	Cups gone.	Cups gone.	Face of Tooth commences to get round.			
25 to 26 Years,	Cups gone.	Cups gone.	Cups gone.	Cups gone.	Round quarter of way down.			
30 Years,	Cups gone.	Cups gone.	Cups gone.	Cups gone.	Round half way down.			

teeth which are gradually replaced by permanent or horse teeth between two and five years of age. The teeth in the upper jaw erupt before those in the lower jaw. In the full or complete mouth the teeth are made up as follows:

The **incisors** comprise the rows of teeth at the front of the mouth and number six on the lower jaw and six on the top. Each row has two corner (outer) teeth, with laterals beside them and two centrals in the middle. The horse relies on his incisors for grazing and biting.

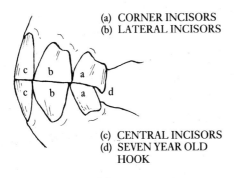

(a) CORNER INCISORS
(b) LATERAL INCISORS

(c) CENTRAL INCISORS
(d) SEVEN YEAR OLD HOOK

Behind the corner incisor teeth on either side of the lower jaw are the **tushes** or tusks. These are sharp, permanent teeth which appear at four to five years of age, usually only in the male horse although mares sometimes grow very small ones. With age the sharp edge becomes rounded off. They do not have any use in the modern horse but can get in the way if the horse accidentally takes hold of a chain or something which might hook onto it.

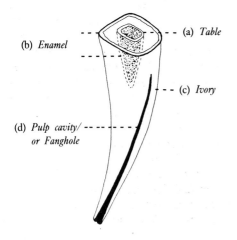

(a) *Table*
(b) *Enamel*
(c) *Ivory*
(d) *Pulp cavity/ or Fanghole*

There are also six **molars** on each side of the upper and lower jaw. The first, second and third are both permanent and temporary but the fourth, fifth and sixth only appear as permanent teeth. The molars or grinders are used for masticating all food.

Wolf teeth, or rudimentary molars, when they grow appear on the upper jaw just in front of the proper molars, one on either side. They do not usually grow more than half an inch in length and serve no purpose at all. Like wisdom teeth in man they can become troublesome and should be removed. Details of this procedure can be found in Chapter 15.

The accompanying illustrations show the normal pattern of the horse's dentition. It can be seen that the semi-circular shape of the incisors gradually becomes straighter with age. Although the upper incisors are in fact longer and larger than the lower ones the

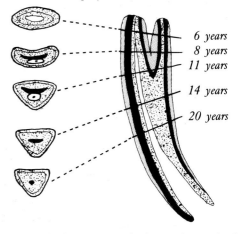

Shape of table of tooth at:

6 years
8 years
11 years
14 years
20 years

structural changes are fairly similar in both jaws. The appearance of the incisors at the age of one year can lead to confusion because the mouth looks similar to that of a five-year-old; however the former has temporary and the latter permanent teeth.

In young horses all incisors and the first three molars are temporary. The rest, i.e. the last three molars at the back of each jaw, the tushes and the wolf teeth are all permanent.

'Bishoping' is the unscrupulous method of disguising the true age of a horse by sawing off the lower incisors until they are level then scooping out a cavity in the corner incisor and blacking it with an iron until it is quite black. The result gives the appearance of a younger mouth although a knowledgeable horseman should have no difficulty in identifying such practice. It is, however, not commonly practised today.

Bone measurement

The horse's bone is measured just below the knee, around the cannon bone, with a tape measure. With practice, however, it is possible to estimate within half an inch by closing a hand around the top of the cannon bone.

The size of a horse's bone is a recognised indication of the weight-bearing capacity. For example, hunters are classified by the weight they are up to carrying, i.e. a lightweight is up to 12 stones 7 lb; a middleweight is between 12 stones 7 lb and 14 stones; and a heavyweight is over 14 stones.

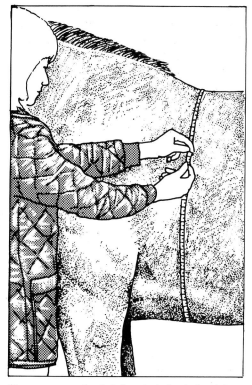

How to measure the girth for calculating bodyweight

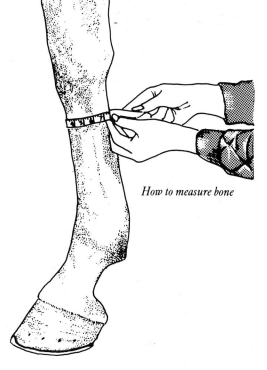

How to measure bone

Girth measurement and bodyweight

The girth is measured for the purpose of determining the horse's bodyweight. An ordinary tape measure or a piece of string will do the job; alternatively one can buy an equitape which both measures the horse's girth and gives you the bodyweight without your having to convert it. When measuring the girth be sure that the tape is placed around the barrel just behind the withers and in front of the girth groove. It is important that the animal is relaxed when you take the measurement which in any case should not be taken just before or after he is fed or exercised. The following table can be used as a guide to converting inches into pounds or kilograms. A public weighbridge can be used to weigh a horse but you must remember to subtract the weight of the tack and handler.

Ponies

Girth (in)		40	42.5	45	47.5	50	52.5	55	57.5				
Girth (cm)		101	108	114	120	127	133	140	146				
Bodyweight (lb)		100	172	235	296	368	430	502	562				
Bodyweight (kg)		45	78	106	134	166	195	227	254				

Horses

Girth (in)	55	57.5	60	62.5	65	67.5	70	72.5	75	77.5	82.5
Girth (cm)	139	146	152	158	165	171	177	184	190	196	209
Bodyweight (lb)	538	613	688	766	851	926	1014	1090	1165	1278	1369
Bodyweight (kg)	244	278	318	347	386	420	459	494	528	579	620

(Tables taken from Glushanok, Rochlitz and Skay, 1981.)

Life expectancy

The horse's life expectancy, can be anything up to thirty-five or forty years and sometimes beyond depending on the work he has endured in his lifetime. The type and intensity of his work is the main cause for limiting his life. It is, however, difficult to judge how long a horse would live if he were not domesticated or, likewise, if he were not subjected to work whilst in captivity. It is significant that the horse's legs will always wear out before his body. Although the old adage says 'no foot, no horse' the limiting factor, from a utilitarian aspect, seems to be the legs, in particular the fore limbs.

No matter what the nature of his work there can be no doubt that the quality of his fitness will contribute most to his overall working life. Each horse's individual capabilities and limitations should be carefully assessed if we are to develop and promote the quality of his life. His physical requirements and psychological or temperamental capacity to withstand the many stressful situations under which man places him should be carefully evaluated and respected and his work programme designed accordingly. If either of these factors is abused by man the quality and duration of a horse's life will suffer. No two horses are alike and where one will be happy to be retired fit and healthy to enjoy the rest of his days at pasture another horse would be miserable and feel abandoned if he has spent his life being stabled and cared for by someone with whon he has established a lifelong relationship.

The length of his working life will depend entirely on how he is cared for, both whilst he is in work and at rest. He should be brought into work slowly and gradually with every care taken to establish a depth to his fitness which will act as a foundation for the work ahead of him.

Death by natural causes does not usually take place quietly. Most horses struggle violently during their last moments although they are unconscious and therefore not in pain. A horse's death in this way is particularly distressing for man, especially for those closely associated with the animal. The chief causes of death in the horse are abdominal pains, such as those which result from colic and related symptoms, and heart failure. More often today the animal meets his end in a humane way either at home or in an abattoir, where the environment and conditions are so designed that the horse has no way of knowing his fate.

BASIC SKILLS IN HORSE MANAGEMENT

Routine observation

One of the secrets of good horsemanship is to be able to sense whether anything is wrong with a horse whenever you are with him. The experienced groom will know instinctively at first sight if a horse is well or not. A natural understanding of the horse's normal disposition together with experience will enable you to decide that something is not right. An abnormality can indicate a minor ailment or a serious disease, and no matter how trivial the symptom, immediate attention will prevent the horse suffering unnecessarily.

A simple way of remembering what to look for generally is to think of the ABC of the horse. Appearance, Behaviour and Condition. Constant observation is therefore a vital part of the everyday routine. Each time you approach a horse be sure you check that he is comfortable in every way. In terms of general health notice whether he is alert with a bright eye and smooth coat. A dull, staring coat is an indication that something is amiss in a stabled animal. He should not be standing with an abnormal stance, for instance with his front and hind legs far apart. This could indicate that he is stretching to ease some abdominal discomfort. He should not hang his head unless he is sleeping. (Don't forget that a horse can sleep standing up.) His tail should not be raised for any length of time unless he is staling or passing dung. Make sure you know what is normal for that particular animal and notice if he is straining uncomfortably.

Each horse is different in temperament and therefore reactions differ but, generally speaking, all horses will acknowledge your approach either by pricking their ears and perhaps neighing or, in the case of bad-tempered animals, by laying back their ears. Any marked changes in temperament could be the result of excitement, fright or pain and should be watched closely in order to diagnose promptly.

To be sure that a horse is warm enough feel his ears regularly in colder weather and after exercise in case he breaks out into a cold sweat. Each time you go into his box make a habit of running your hands down his legs. Only familiarity with each horse's conformation will allow you to recognise any changes and deal with them promptly.

Swelling in both hind legs can sometimes reflect a combination of digestive troubles and insufficient exercise. To rest a horse or reduce his exercise without adjusting the feed accordingly often predisposes swelling of the legs particularly in the hind limbs. Older horses are more prone to filled legs and in some cases the problem refuses to improve or does so only after exercise. If only one leg is filled an injury to that limb should be diagnosed before work is resumed. The day after a competition or hunting is often when wounds, perhaps caused by thorns, will manifest themselves, so a thorough examination of the limbs should be carried out before an infection develops. Any thickening of a tendon should be regarded seriously and remedial action taken because to work a sprained limb is to jeopardise the soundness of the horse in the long term.

The smallest swelling or presence of heat requires early detection if it is to be given every chance of a speedy recovery with correct treatment. The old adage of 'a stitch in time saves nine' is never more true than with horses' limbs.

When you visit a horse in his stable his breathing should be that of a horse at rest, i.e. ten to twelve breaths per minute without any noise or physical contractions. Any discharge from the nose should be regarded as abnormal and an indication of possible illness. The skin of a healthy animal feels soft and supple to the touch and if he is in good condition there will be a bloom to the coat. The mucous membranes of the nostrils and eyes should be pink in colour. The normal pulse of a horse at rest is between thirty-six and forty beats per minute. His temperature under normal conditions would be 99.8–101°F (around 38°C). Attention should also be paid to the animal's droppings because there can be a number of reasons for any abnormalities other than a change of diet or excitement.

Any sign of a runny nose or eyes, a cough or a respiratory noise of any sort should be attended to at once and the horse's work routine adjusted accordingly. There is a very real danger of permanent damage to the heart and lungs if the horse is overstressed through work or travel when unwell, so early symptoms should not be ignored. It is advisable to seek veterinary advice at once if the horse has a temperature.

If the horse is found to be rolling a great deal this is often an indication of an abdominal disorder, such as colic, and should be dealt with immediately. Look at the animal's flanks to see whether he is relaxed or not. If he is tucked up find out why; it may be because he is cold but it could also be a sign of ill-health. Pawing the ground for any length of time is unnatural and may indicate colic. The horse will probably dig up his bed before he rolls and you will have to decide whether his behaviour is normal or if he is showing signs of discomfort.

The feed manger and water container should be checked regularly to ensure that the animal is eating up, particularly his late night

A horse in good condition.

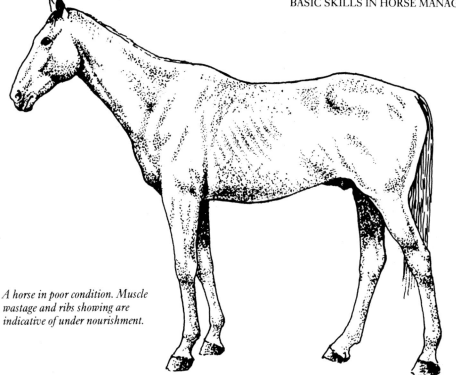

A horse in poor condition. Muscle wastage and ribs showing are indicative of under nourishment.

feed which may have additives and supplements in it. It is not uncommon and not necessarily a sign of ill-health if the horse does not eat up his daytime feeds straight away, especially once he is in work. The horse's water intake should be monitored, which is difficult if automatic water bowls are used. Keep a check also on the amount of hay he eats. It is uncommon for a horse to leave his hay ration even though he may be off his other feed.

Routine observation will, with practice, become second nature to you but each horse is different and must be treated as an individual. It is therefore important to establish an intimate knowledge of each horse and learn what is normal for him.

Signs of good and bad condition

In good condition the horse will be seen to be not only surviving but also thriving. His coat should be smooth and have a bloom to it. He should be well covered with flesh, particularly over his quarters and neck, and the skin covering his ribs should be loose. When horses are living out in winter their looks can often be deceptive and the experienced eye will have learnt to recognise when a thick winter coat is disguising a poor condition. Another time for caution is in springtime when the animal is shedding his winter coat and growing a summer one. Care should be taken that the horse does not catch a cough or chill especially if the weather is wet and the atmosphere constantly damp. Horses can happily tolerate cold weather but will not stand wet and cold at the same time. A large 'pot' belly can falsely represent a fat horse in good condition and may be an indication of worms, quite common in youngstock.

Having established the points which constitute a good condition in the horse it follows that the horseman should be able to identify immediately the signs of poor or bad condition. If the animal is underfed and therefore undernourished his coat will be dull and there will be an apparent weight loss with the animal's ribs showing and possibly his hip bones and withers protruding. His neck and rump will be wasted of flesh and his vertebrae may be prominent. The horse's general demeanour will be lethargic and his vital signs will be abnormal in extreme cases. There will be an unwillingness to exercise himself or allow himself to be exercised. His movements may be laboured and his action might appear lazy. A lack of vitality together with a dull eye and uninterested expression are further indications of bad condition. In extreme cases subclinical symptoms may become apparent, such as a discharge from the eyes or nostrils, digestive disorders and abnormalities in the dung and urine. His stance will be unnatural, perhaps with his tail raised for no obvious reason.

Stable manners

Good stable manners in the horse are an essential aspect of his handling and management. A well-trained horse is both safer and a pleasure to own. The horse must learn from the start that a mutual confidence, trust and respect are the basic principles on which the relationship with his handler is built. Without respect for his handler bad stable manners and habits will ensue and the horse will become ignorant and unruly; no matter how talented a performer he may be, he will be unpleasant to care for. The safety factor involved is often underestimated by the novice horseman – a grave mistake because both the horse and his handler, as well as any assistants, are at risk not to mention members of the public. The handler should always be firm and instil discipline into the horse from an early age, using the voice to educate. The sooner this is established the better, beginning when the animal is still a foal, teaching it to develop confidence at the same time. So much of the ignorance and bad manners seen in the horse is a direct result of the owner's mishandling and mismanagement, mistakenly being too kind to the horse. For example, being over-generous with titbits may encourage the horse to take advantage. Correcting other people's mistakes through re-education can often be slow, dif-

ficult, counter-productive and sometim impossible.

Many horses will change their attitu towards man at feed time and often a normal well-behaved animal will lay back his ears a threaten the feeder on his way to the mang Again, firmness is needed, and from an ea stage, to teach the horse to wait until you h placed the feed in the manger. A horse t is deliberately rude to his handler should punished promptly so that he connects y reaction with his action and thereby le right from wrong. It must be stressed here no matter what form of punishment choose he should NEVER be struck abou head. The horse that becomes head-sl a result of being hit on the head is diffic handle and can, depending on his tem ment, develop other habits. He will co quently be suspicious of anything you wish do around his head.

The horse should be taught from the sta that whenever you are with him he shoul move over promptly when asked, and that h should stand to one side until you move him This should be taught to him when he is tie up and preferably at an early age, so that onc the command is established he will alway obey you promptly. The horse must learn tc react willingly and spontaneously to you requests.

Good manners in the horse are the found tion of any relationship between horse an man and should never be allowed to slip because the horse will be quick to take advan tage both in and out of the stable. Apart from the points already mentioned a rude horse ca be a source of embarrassment and humiliatic to his handler.

Approaching a horse in the stable

Always announce your arrival before you open the stable door. Never surprise a horse by approaching him without speaking to him first and holding out your hand. The horse who is otherwise of a quiet temperament may react violently, i.e. by kicking, if he is startled and alarmed. It is the horse's natural instinct to be defensive so it is important that you approach in a calm but direct manner. He should always be encouraged to come towards you. It is not necessary to offer the horse a tit-bit each time you see him. The exception to this is the young or timid animal who lacks confidence in man; in this case a bowl or bucket containing a few oats or nuts will encourage him to be handled. As soon as the animal has overcome his fear it is advisable to dispense with the bait before he comes to rely on it.

Whenever you visit the horse and find him standing by the door, insist that he stands back whilst you enter. He should also be taught to stand up and be still whilst you are in the box. When working around him in the stable, e.g. when mucking out, grooming, tacking-up, rugging, changing rugs and bandages etc., it is recommended that he is tied up by a headcollar and rope to a fixed ring in the wall. Always close the stable door after you, even if you know the horse well and are confident that he will not pass you once you are inside. Never be left to close the stable door after the horse has bolted!

Handling different horses according to age and temperament

Confidence in handling horses at all times is a prerequisite in terms of establishing respect, trust and confidence from the animal. He should be dealt with in a direct and positive manner at all times with an awareness for safety constantly in mind. Anticipation of an animal's reaction to a given situation is born out of experience of horses and an intimate knowledge of each horse. Familiarity with an individual can lead to complacency and over-confidence, which is often why accidents occur.

Through his superior strength the horse could always have an advantage over man if ever he wanted and this is something the handler should respect. Old-fashioned ways of taming the horse, forcing him to submit through the use of rough and uncultured methods, are thankfully a thing of the past.

Patience, intelligence and a sincere affection for the horse are some of the necessary attributes of the handler/trainer. A good horseman has a quiet but firm attitude to his horses, only raising his voice to reprimand, praising whenever necessary, anticipating trouble and dealing with it promptly. If the horse is to learn he must be given every chance to understand what is required of him, be encouraged to cooperate and kept calm without fear or excitement of a situation or person. Bad manners on the part of the horse can be the result of his misunderstanding your instructions, so be sure not to confuse or alarm him for he is easily excitable.

The horse's age, temperament and character should be given due consideration at all times. Whilst each horse's courage and intelligence vary and therefore his capacity for learning, the trainer must ensure that his pupil is never asked too much too soon. Each lesson should be thoroughly established before the next is introduced. Do not expect a fresh horse to absorb any training until he is settled and attentive. He will not concentrate on his work whilst his attention is diverted through high spirits or other distractions. At the same time he should not be allowed to take advantage; the observant handler/trainer will recognise this. It is a mistake to expect the same response from a horse two days running. His education should be regarded as rungs on a ladder. If his progress is not kept within his capabilities he should be taken back to a lower 'rung' until he is ready to move up again. A horse with above-average confidence and intelligence should not be taken too fast or his generosity abused; sooner or later he will come unstuck and you may have to go right back to basics. Always be sure that the horse understands what is required of him and reward him accordingly.

Use of headcollars, halters and tying up

Correct fitting of the headcollar or halter is essential for the sake of the horse's comfort and safety and for your control over him. There are three sizes: pony, cob and full-size, and some can be adjusted to fit. Leather or nylon headcollars should never be used if they appear worn or with loose stitching, buckles or rings. The same applies to ropes which are used for tying up because they are easily broken. Rack chains are an alternative to a rope for tying up, but they can be dangerous. Great care should be taken that the horse does not take the chain into his mouth, and that the length is suitable so he cannot get his leg over it or put his head underneath it, hooking himself up. Never leave a rack chain in the box when it is not in use or lead a horse with it. Less expensive alternatives, such as plaited baler twine, are a saving if a horse is known to chew a rope and most young horses do so at some stage.

While halters may be much less expensive the bull type is not recommended for tying up a horse. The rope which passes under the jaw can tighten onto itself and possibly injure the horse who may, as a consequence, become head-shy. The Yorkshire halter, which is made of webbing and is non-adjustable (unlike the bull halter), is safer to use for tying up because a rope or chain can be attached to it.

The principal thing to remember when tying up a horse is always tie the animal to an immovable fixture, never to anything which could move if he pulled back. A tie-ring should be cemented into the wall, no less than 5 ft (150 cm) from the ground. Below that height there is a danger that the horse could

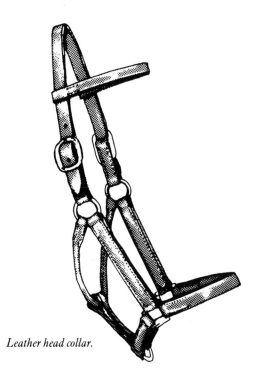

Leather head collar.

catch his leg over the rope. Secondly, to avoid the risk of the horse breaking a headcollar, halter or rope, if he pulls back, be sure to have a short piece of string (a loop) attached to the ring and fasten the rope or chain to this *not* to the ring. Never tie a horse so that he can graze, because of the risk of him getting his foot over the rope.

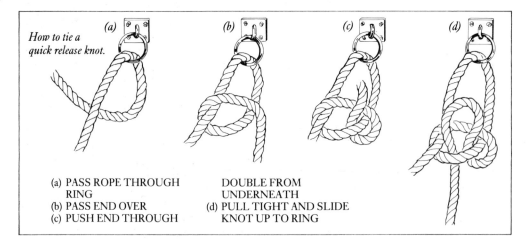

How to tie a quick release knot.

(a) PASS ROPE THROUGH RING
(b) PASS END OVER
(c) PUSH END THROUGH

DOUBLE FROM UNDERNEATH
(d) PULL TIGHT AND SLIDE KNOT UP TO RING

A horse should first be taught to accept being tied up in a stable, preferably by two people, one of whom should be at his head and the other behind him teaching him to accept the restriction and not pull back. He can also be taught to move over from side to side whilst he is tied. Time spent on this part of his education is an investment for once he has learnt properly he will know what is expected of him. If at this stage he learns to pull back and break loose he will always try it and thus become a nuisance.

When approaching a horse that is tied up be sure he knows that you are there; do not risk walking straight up to his quarters and patting him. Let him see you at his side first and reassure him before you do anything to him.

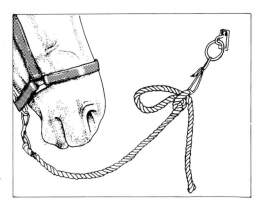

Tying up a horse.

Leading in-hand and trotting up for inspection

In stable management there are several golden rules to learn, some of which come under this heading. First of all, whenever a horse is lead out of his stable the handler should ensure that the headcollar and lead rope are adequate, i.e. they must fit the horse, be sound and correctly fastened to allow the handler as much control as possible. Never take a young, fresh or strange horse out wearing anything less than a bridle (without passing the reins over the horse's head) or a strong headcollar with a lunge line attached. A short lead rope does not allow for any misbehaviour and the horse can soon get away. Once he learns that he can do this there is always a danger that he will try it again. Never lead a horse with a chain and on no account wrap a lead rope or rein around your hand. The slightest pull by the horse could injure your wrist or worse. Gloves should be worn to protect the hands from rope burns.

Any attachment which is used to lead the horse, e.g. rein, rope and clip, should be sound and strong. Ropes that have been chewed or rusty clips are not good enough. Particular care should be taken when leading in and out of stables and through gateways. Always be sure that the horse is straight and is not in danger of catching his hip if he has to turn. The door, gate or slip rails should be opened fully and no short cuts taken. If necessary, fasten them back so that they do not swing back as the horse is passing through. A young horse can be easily startled by a door or gate blown back by the wind just after he has walked through, and there is a danger here that he will attempt to rush as a result of this one bad experience. If there are horses being led in front or behind you ensure that there is at least a horse's length between the animals.

The horse should be led from his near side, i.e. with him on your right. Your right hand should hold the lead rein near to the headcollar (depending on the amount of control needed) with the rest of the rein in your left hand. At all times he should be slightly in front of you so that your shoulder is level with his. Movement at whatever pace should be purposeful and straight to encourage the horse to exhibit himself well. Having a horse that hangs back from his handler and makes no effort to move forward is both inconvenient and suggests an idle attitude on the part of the handler. A young horse should be led with an assistant walking behind him until he learns what is expected. A lazy horse should be encouraged with the help of a long enough whip, if used promptly, will soon teach him to be active. The handler should avoid looking back at the horse because this often discourages the latter from going forward.

How not to lead a horse.

How not to lead a horse.

Leading in-hand is one of the most important and basic aspects of equine management. Throughout its life the horse will be required to trot up for soundness, sale and show so it is essential that the handler knows the correct procedure. First, be sure to take the horse onto a straight, level stretch of hard surface, e.g. a driveway or quiet road which is at least 40 yards (35 m) long. He must have enough room to trot out and pull up again in a straight line and to turn *away* from you (never towards you) on not too sharp an arc (allowing for unsoundness especially in front limbs). Quiet surroundings encourage the horse to relax and show himself more naturally. If an unsound horse is distracted or alarmed when he is trotted up he may give a false impression. The horse should be brought from his box and asked to stand straight and square for inspection with the groom standing in front facing him. Next he should be walked away from his examiner(s) for 35–40 yards (30–35 m) or longer if required, then turned round and walked back past the person(s) watching him. He will then be expected to repeat this at the trot (unless, of course, he is too unsound to trot). The first four or five strides should be at the walk to give the horse a chance to balance himself before springing into trot calmly. Likewise, after he has turned at the far end, wait until he is straight and quiet before asking him to trot back again. Do not pull him up suddenly, hold his head too tight or turn his head towards you as this can disguise a 'nodding' head – an indication of lameness.

One final point on this subject but one which applies equally throughout stable management, is the question of the handler's footwear. Although it is often disregarded it is an important safety factor because no one can trot up a horse properly with sloppy or high-heeled shoes. There is also the danger of being trodden on by the horse.

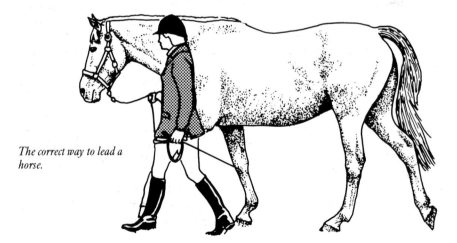

The correct way to lead a horse.

Catching a loose horse

The prime concern when catching a loose horse or horses must be one of safety. Any outside gates should be fastened immediately to try and contain the horse in as small a space as possible. Assistance should be sought wherever possible and any obvious mobile hazards, such as a wheelbarrow parked in the open, removed because an excited horse will invariably run into it. At least one headcollar should be available together with a bucket containing some feed. Approach the horse slowly and quietly whilst offering him the feed. Do not hurry in any way and be sure that any loose dogs or children are out of the way. A galloping horse can very quickly do a lot of damage and even the quietest of temperaments can become excited. Aim to confine the animal in a corner, preferably with one person either side, each with a headcollar. Never try to herd him or hurry him because he will invariably go faster and a loose horse will often behave wildly with no respect for fixtures and fittings. If possible, have someone bring out another, sensible horse (wearing a bridle or headcollar and lunge line in case he too becomes excited) because often a loose horse will follow another of his kind back to the stables. Once the horse is willing to be caught be sure to restrain him firstly by putting a rope around his neck before fitting the headcollar because he will have enjoyed his moments of freedom and may take it into his head to try and get loose again if he thinks he may have a chance.

If the loose horse is tacked up and you are out riding, do not be tempted to chase after him because invariably he will go faster. Wait until he stops and it is safe for you to ride up alongside and get hold of the reins and take them over his head. It may be that the horse you are riding will not be willing for you to lead the other horse, so you may have to dismount and lead them both. If the runaway has lost his bridle use the thong of your hunting whip (or even your neck tie) around the top of his neck to restrain him until a headcollar or bridle becomes available.

A cast horse and how to deal with him

A cast horse is one which, having lain down and tried to roll over, has become stuck on his back and cannot right himself. He may have rolled and become cast due to an attack of colic, in which case the vet should be summoned immediately if the animal is struggling and in pain. Colic can result in a twisted gut, which is usually fatal.

If he becomes wedged against a manger or a door it is often more difficult to free him, but in any event a procedure must be adopted in the interests of safety. When a cast horse is first discovered help should be summoned immediately. Never try to get a horse up again on your own because cast horses often flail their legs about in an attempt to regain their balance, which is dangerous for anyone nearby. Once help is found someone should approach the horse and sit on its head, using all his weight to restrain the horse. Thick ropes, such as plough lines or lunge lines, should be attached to the hind and fore legs which are underneath and the horse pulled towards the middle of the box. Righting a cast horse is a potentially hazardous event particularly if the animal panics. It is therefore essential that the utmost caution is taken and that helpers are aware of the danger.

One of the precautions against a horse getting cast is to fit an anti-cast roller, although it must be said that if a horse does manage to get cast whilst wearing one it is virtually impossible to pull him back without first cutting the roller to remove it. This is extremely difficult to do when the weight of the animal is against it. The various types of anti-cast rollers are described in Chapter 11.

Once the horse is on his feet again examine him thoroughly for any injuries or wounds, such as grazing of the joints, particularly the hocks and knees. There is a danger of a back injury, which should be recognised and if necessary professional advice sought. If the horse rolls whilst you are with him and appears to be getting cast you may be able to get him to his feet again by startling him, perhaps by clapping your hands or banging a bucket.

A horse cast under the manger.

How to extricate a cast horse.

Handling a horse for treatment and clipping

An assistant is often required to hold a horse whilst he is being given veterinary treatment or being clipped, so the helper must know what to do. Never allow a nervous assistant to handle a horse in these situations. In the interests of safety choose someone with an alert mind and an awareness of the potential dangers involved with horses who may be in pain. Some incidents can be avoided if the handler appreciates how a horse may react, but accidents cannot always be prevented.

If a horse's leg has to be held up, as a means of gentle restraint, the assistant should be on the same side as the leg he is holding. If he is holding the rope at the same time he should guard against the horse snatching his leg and getting it over the rope. The assistant will be relied upon to hold onto the horse for as long as is necessary and not to let go without warning otherwise he could endanger the safety of people who are around the horse.

It is important to know the horse's temperament and be familiar with his mannerisms, likes and dislikes and how he is likely to react to certain situations. For example, he may be difficult to clip around the head or may resent having his legs hosed. Some horses are sufficiently distracted by eating a small feed at the same time; others, however, do not respond to quiet, patient handling and have to be restrained more severely. Depending on the animal's temperament a headcollar and long rope may offer sufficient control; other horses need a bridle. In order to effect a treatment it may be helpful to back the horse into a corner. If he struggles and tries to evade his handler or snatches to free his leg from the assistant's grip, it will be up to that person to hold on and not give in to the horse. Often he will submit once he realises his handlers are determined.

Awkward horses, that deliberately misbehave, will often need more severe restraints at the outset. If the horse is a known kicker a front leg can be tied up with a stirrup leather, folding it once round the hoof, back round the forearm and buckling it tight. A knee cap should be fitted as a safeguard should the horse try to go down.

Perhaps the most common method of subjugation is with the twitch. It is, however, vitally important that it is used correctly otherwise accidents and often irreparable damage (usually to the horse's behaviour) can happen.

A twitch can be made of a piece of round wood similar to a broom handle, 2–5 ft (60–150 cm) long. A length of strong cord, about the thickness of a small finger, 1–2 ft (30–60 cm) long, is then passed through a small hole made near the end of the stick and tied to form a continuous loop. (Window cord is ideal but *never* use baling string.) The loop is first passed over the hand and the twitch applied by grasping the horse's top lip, sliding the loop over the hand so that the twitch can be twisted to tighten the loop around the horse's lip causing him a certain amount of discomfort. It is important that the tightening up and slackening off is done promptly because the horse may fight against it and once he breaks free from its grip he will always know how. The tension can be adjusted according to the horse's response.

An alternative to the home-made wooden twitch is the American metal version which is shaped like a large nutcracker with a piece of cord or leather at the open end which fastens it together when it is closed around the horse's lip. The advantage of this type is that once it is fitted it holds in place by itself whereas the wooden stick type has to be held by someone. It is not recommended that the American type be used by novices or on young or temperamental horses because if the horse took fright it might be difficult to catch him to release it.

Another method of subduing a horse is to take hold of an ear but not every horse will respond to this and some may become nervous about their ears being touched. Sometimes pinching a fold of skin in the middle of the horse's neck will cause him to lower his head and behave submissively. In extremely difficult cases it may be necessary to administer a sedative before the horse can be treated or clipped which will involve a visit from the vet.

CHAPTER THREE

PRINCIPLES OF STABLE MANAGEMENT

Day-to-day management

The operation of any yard relies on an efficient and organised working staff who between them carry out a timetable which is designed to give priority to the horse's welfare, i.e. his exercise, feeding, grooming and detailed routine for which he relies totally on man. In the past it was always the custom of a stable yard to employ a male stud groom who did not necessarily ride but was totally responsible to the owner for the overall day-to-day management of the yard and its residents. Nowadays his role is taken by a head girl or head lad and it is more commonplace to find a stud groom in a racing or stud yard.

The number of staff required in a stable yard will depend on the number of animals to be cared for and the work they are doing. Principles have changed over the years and where it was once considered essential that no groom should care for more than two horses, a groom today may be found 'doing' any number, especially if they are not in work. It is, however, recommended that one groom should have no more than three working horses in his charge as more often than not the modern groom is expected to ride out.

It is the duty of the head person or manager to allot horses to grooms depending on their experience, temperament and ability. For example, a novice teenager should not be expected to care for a temperamental or nervous youngster without supervision. The knowledgeable horseman will pair up horses and grooms to the advantage of both and recognise the need for change before it is too late. An apprentice or trainee should be under constant supervision and not be expected to know all the details of the job which are second nature to an experienced groom. The head person must expect to bear the full responsibilities which go with his position and to make quick decisions in an emergency. It is important for him to understand 'man management' and develop respect and loyalty from his staff because the smooth running of the yard relies on team effort and the full cooperation of all concerned. The atmosphere of the yard reflects the mood of the staff and subsequently affects the behaviour of the horses. They will not generally relax whilst there is activity and noise on the yard.

As well as organising the staff the head person will be expected to liaise with the owner and plan and execute a timetable for the horses' work each day. He will also be expected to know how to adjust each horse's programme for reasons such as the animal's health, the weather and ground conditions. It will be up to him to pair up horse and rider appropriately and to make changes if a horse and rider are not compatible. Feeding, ordering of feedstuffs, bedding materials and stable equipment and overseeing the maintenance of these items will be his responsibility too. Treatment and clipping is often carried out by the head person where it is not practical to delegate. The trainee on the other hand, is engaged in the yard to learn his trade and it is up to the head person to ensure that this training is carried out.

In a stable yard where one groom is employed and is directly responsible to the owner it will be assumed that this person is sufficiently experienced to undertake the day-to-day management if the owner is not actively involved with everyday duties. It is inappropriate for the owner to expect a novice groom to be given the entire responsibilities involved with the welfare and exercise of valuable horses.

The role of the groom

Before outlining the duties of the groom it is appropriate to suggest the qualities which an employer would look for when hiring him. The groom is understood to have a knowledge of the horse and how to care for him. He is referred to as a groom as opposed to a trainee/apprentice/working pupil, and should be regarded as having an entirely different status. His experience will reflect his capabilities according to his age and the length of time he has been employed in a stable. The good groom will have an even temper, a confident manner with his charges, be entirely dedicated to their welfare and be a conscientious worker who takes pride in his horses and their surroundings. He should be prepared to carry out his duties in all weathers and not be a clock watcher. He should have a good constitution and be physically fit and strong. In competition yards the groom will have to learn to become adaptable to the flexible schedule which events invoke. Long and irregular hours are part of the day-to-day routine once the competition season begins. It is therefore imperative that the groom should be prepared both physically and mentally for the constant pressures of such a lifestyle. It will also require great tolerance to withstand the tension of competitions for the rider's anxiety can often rub off on the groom. Because of the irregular schedule during the season the groom may be expected to adjust his timetable with regard to time off and holidays. It should be a matter of give and take, making the most of slack periods and being available during busier times.

The qualities of a groom are suitably summed up in Mayhew's *Illustrated Horse Management* (1891) as follows:

"Let the groom from the earliest moment of service, make up his mind to serve his master with truth and honesty and to avoid the company of the reckless and ignorant and to seek advice from those who are competent to give it; to treat with kindness and consideration the lives of the horses committed to his charge; to use kind words instead of harsh expressions; to educate the animals to fondle rather than to fear, and, in the long run, and short too, you will discover, O, groom! that, with these three weapons, – truth, decision, and kindness, – you will be able to direct horses almost with your hand."

The duties of the groom

The duties of the groom will largely depend on his experience. The trainee school-leaver should not be left to do anything he has not been properly trained to do and must be subsequently monitored until it is found that he is capable of carrying out that task. The skilled groom, however, will be expected to perform most stable duties singlehanded and efficiently.

Generally the first job of the day is feeding and this is usually supervised by a senior member of staff or the owner. It is up to the groom to check that the horses he feeds have eaten up the last feed of the previous day and to report back to his superior as to how the horses are. Mucking out and yard sweeping is normally done before breakfast and it is up to each groom to ensure that his horses are done properly and on time. Slow workers should allow longer for chores such as mucking out but at the same time learn by experience how to improve their efficiency. Breakfast is usually taken when the stables are finished, after which exercising begins. Those yards which prefer their horses ridden before breakfast may expect the groom to tack them up before or during mucking out, so enough time should be allowed for this even if it means the groom getting up earlier.

During breakfast, or in some yards the night

41

Person dressed correctly for work in the stable yard (hair tied back and wearing sensible footwear).

before, the riding plans are decided so each groom will know which horses he is to ride and which he is expected to prepare for someone else to ride. The head girl/lad shall ensure that each groom knows exactly what tack is used for each horse and how it is fitted. If there is a change in the horse's regular tack it is up to the head person to make sure that the relevant groom knows how to fit it. The groom may be expected to bring the saddled horse out to be mounted at a specific time and must ensure that he is punctual. Courtesy to the owner should be expressed at all times, not least when the horse is presented to him. Once the horse returns from exercise the groom should care for him promptly by washing down or grooming. The lunch-time feed then precedes the staff's lunch break after which some yards insist on the horses having their *siesta*. At this time the grooms are usually employed on tack cleaning and carrying out maintenance chores.

Afternoon stables involve the groom any grooming, strapping, clipping, trim etc before the horses are set fair for the The groom sets fair the beds of his ho gives hay and water, changes rugs and up before feed time. Any veterinary t is done at this time too and the gro have to carry this out himself or assist s else. The groom checks his horses ag thing, i.e. between 9 and 10 pm, makin that water buckets are full, rugs straigh that the horse has finished his tea-time 1 Where appropriate the groom may gi fourth feed before leaving the horse fo night.

Example routines for different types (

Racing yard

0600 hours	Feed – head lad only.
0700	Lads arrive on yard.
	Muck out, leave bed 1 over horses, put in water, tack up – fi (group of horses) onl
0745	First string exercised.
0915	First string return to yard, brush over, pick out f up, set bed fair, feed.
0930	Lads go to breakfast.
1000	Lads return to yard.
	Prepare second string.
1045	Second string exercised.
1200	Second string return to yard over etc and fed.
1230	Half the lads ride out the string, and the other lads those sick or lame animal: in the yard.
	Remaining lads muck ou third string and prepare feeds.
1315	Third string return to yard, over and fed.
1330	Lads to lunch.
1600	Lads return to yard, skep ou horses and groom.
1700	Trainer's inspection.

1715	Once inspected rug up, hay and water.
1745	Feed.
1800	Finish afternoon stables.
2100	Night check and fourth feed as necessary.

jumping yard

0700	Staff arrive on yard; feed.
0715	Muck out all horses, sweep yard, dress over horses.
0800	Staff to breakfast.
0845	Staff return to yard, prepare first lot of horses for work.
0900	First horses worked.
1015	First horses return to yard, untack and brush over or wash off.
1030	Second lot of horses prepared for work.
1040	Second lot of horses worked.
1145	Second lot of horses return to yard, untack and dress over. Remaining horses exercised. All horses skepped out and fed, staff to lunch. Staff return to yard, grooming, trimming, tack cleaning, veterinary treatments and any other chores carried out, check horses' soundness where in doubt.
1630	Doing up time – skepping out, water, hay, rugs changed, yards swept.
1700	Feed time.
1730	Staff leave yard.
1800	Evening stables including fourth feed where applicable.

Daily chores extra to horse routine

1. Muck heap.
2. Yard sweeping, raking, weeding.
3. Water containers and feed mangers to clean.
4. Checking horses at grass, paddocks, water troughs and fencing.
5. Maintenance jobs.

Weekly chores

Monday
Clean stable windows and remove cobwebs.
Sort out any problems arisen over the weekend.
Check fields, fencing and troughs.
Running repairs, vehicle check.
Rake in edges of manège.

Tuesday
Clean out all drains and drain trap.
Check first aid and fire-fighting kits.
Check light bulbs.
Check all equipment is in proper place.

Wednesday
Clean out tack room.
Clean and check surplus tack.
Clean grooming kits.
Clean headcollars.
Repairs to bandages and rugs.
Brush out rugs.

Thursday
Clean out feed room, hay and straw sheds.
Check feed stocks.
Order feeds, fetch any orders.
Scrub out feed and water containers.
Harrow indoor school.

Friday
Clean out cloakroom.
Clean out ancillary buildings, e.g. office, indoor school gallery, store sheds.
Check all horses' shoes/feet and arrange blacksmith visit as necessary.
Check stable diary and records for due dates re worming, vaccinations, any treatments.
Check staff rota for weekend.

Routine seasonal work

Horses:
Dental.
Worming.
Tetanus vaccinations.
Flu vaccinations.
Clipping and trimming.
Getting up and roughing off.
Work.
Competitions.
Monthly foot trim for horses at grass.

Preparations for mating, foaling, weaning, culling and sales.

Stables:
Gutters and roof.
Electrical safety.
Plumbing.
Painting and pointing.
Maintenance and repairs, doors to rehang.
Spring clean and disinfect.
Rodent control.
Timber preservation.
Clean out store sheds.
Check fire appliances.

Property:
Fencing maintenance.
Roads and tracks.
Woodland.
Hay racks.
Gallops.
Arena or manège.
Notice boards.
Water troughs.
Hedging and ditching.
Field drainage.
Gates and gateways.
Check transport and any machinery, MOT etc.

Grassland:
Rotational and alternative grazing.
Harrow droppings on hot, dry days or pick up whenever practicable.
Harrow dead grass.
Roll as necessary.
Soil analysis.
Fertilise.
Mowing.
Kill weeds using knapsack sprayer.

Management Activities:
Book-keeping; accounts, VAT returns, wages (PAYE), budgets etc.

Stable diary

Whatever the size of yard in terms of number of horses and staff a decent-sized stable diary is essential. Every aspect of the yard management should be recorded and it should be one person's responsibility to ensure that it is kept up to date every day. Some yards prefer to keep a separate book for shoeing and another for veterinary treatments but you may prefer one book for everything. The diary should contain details of every horse such as:

when he came into work or arrived in yard;
his training schedule and details of day's work;
competition programme and results;
his holidays;
his rest day;
any vices or peculiarities;
veterinary treatments;
blacksmith's visits and shoeing req ments;
feed chart with notes of changes mad
vaccination schedule for equine infl and tetanus;
worming dates;
allergies and reactions to the same.

Other details which may be included are
staff rota including time off and holi arrangements;
client's or owner's riding plans;
appointments and visits;
vehicle servicing/MOT/repairs including horsebox or trailer;
accident record (if a separate accident book is not kept);
telephone numbers and addresses of all personnel connected with the yard, i.e. vet, blacksmith, fodder and bedding merchants, staff, clients, owners, proprietors, emergency services, e.g. local police, fire brigade, doctor, hospital, insurance company;
list of competitions, hunting dates;
saddlery repairs (particularly important in livery yards).

Accident book

The Health and Safety Act 1974 requires that any establishment which employs personnel must keep a record of any accidents that occur during the employees' employment. It should be made available to the Health and Safety Officer whenever he visits. The details must include when and where the accident took place, who and what was involved, the names of any witnesses, and injuries, hospital treatment or otherwise.

Suitable clothing for work with horses

The nature of stable work dictates that practical clothing must be worn. Safety must always be given prime consideration and whilst cost may be a priority for many people it may be false economy to buy cheap footwear or riding wear.

Footwear

For stable duties off the horse a good strong pair of shoes, boots or wellingtons must be chosen in preference to light gym shoes or shoes with big heels, not only because there is always a danger of the groom being trodden on but footing must be secure to give adequate purchase when handling a horse. For riding only correct footwear such as jodhpur boots, rubber or leather long boots should be used. On no account should shoes or wellingtons be worn, for no matter how competent you are in the saddle your foot can easily slip through the stirrup and endanger your safety. Footwear is soon affected by the urine in the stable and will rot away if it is not cared for by washing and polishing regularly.

Breeches/jodhpurs

There is a wide choice of breeches and jodhpurs available today and in far more practical colours than the conventional cream or white which soon get dirty in the stable yard. A pair of navy blue, brown or black jodhpurs will look better for longer but a pair of cream breeches or jodhpurs should still be kept for competitions etc.

Coats/jackets

An anorak for everyday use will be an essential part of your wardrobe. It will get a lot of wear and tear and will therefore need to be of good quality if it is to last for any length of time. When buying an anorak be sure it is warm enough because you will have to ride in it in all weathers. To save your anorak and protect other clothes a smock can be particularly useful. Alternatively a boiler-/jump-suit type of overall will protect you when doing the dirtier jobs such as clipping. It may be necessary to own a hacking jacket for wearing at shows and while they are normally expensive it is often possible to find good secondhand ones in riding wear and saddlery shops. Waterproof coats and leggings are particularly useful and here again false economy should be avoided. Quality waxed and thornproof materials are preferred to nylon, which is easily torn and tends to cause sweating underneath. The traditional cream riding mackintosh is made of superior quality material but its price does put it out of reach for many.

Hats

For the purposes of riding a hard hat or crash cap must always be worn. This may become compulsory by law in the course of time because riding has more casualties with head injuries than any other sport. This is one area where expense should not be spared at the risk of one's life. Be sure that the size is correct and that the hat complies with British Standards Institution regulations. Around the yard it is recommended that long hair should be tied back so that it does not hang over the face. It is not only unhygienic but also a safety risk – the hair could get caught up in grooming or clipping machinery. Headscarves look tidy and are suitable headwear.

There is no reason why, given the range of outdoor and riding clothing available today, the groom should allow him/herself to become dirty and untidy at work. An untidy groom reflects his work and environment and is an embarrassment to his colleagues. Although it is acknowledged that the work can be dirty the groom should allow for this and wear protective clothing, keep him/herself clean and change often. For the purposes of competition and whilst attending to horses away from home the groom should keep a spare outfit of breeches or jodhpurs, trousers, shirts, shoes, and jacket or coat because there is nothing more unsightly than a scruffy groom.

Stable hygiene

The general appearance of the stable yard reflects the hygiene and subsequently the general welfare of the horses. The standard of stable management can often be recognised by the state of the tack room and the muck heap. Poor stable management, careless and untidy staff who have no pride in their work will have a detrimental effect on the entire environment resulting in poor hygiene. A poor level of hygiene endangers the horses' health and can lead to contagious diseases and infectious coughs and colds. Vermin are a constant hazard in the stable yard, particularly around feed stores, and can be discouraged if the area is kept clean and some sort of vermin control is enforced. There is no point in paying for vermin control if the housekeeping is disregarded.

Any cleaning job is made easier if it is done regularly. Cleaning, washing and scrubbing of equipment such as the tools which are themselves used for cleaning should be part of the day-to-day routine. Broom handles and grooming kits should be cleaned at least once a week by scrubbing in disinfected water or in hot soda water. Feeding receptacles, mangers and water bowls should be scrubbed weekly and stale food removed daily. Removable mangers can be easily washed under the tap but others may have to be soaked in-situ in which case it may be prudent to wash these more often to prevent a build-up of food particles and grease. Attention should be paid to keeping drains and gutters clean and germ-free. During the summer it is good practice to wash the stables down, perhaps as much as once a month wherever practicable, with a liberal application of disinfectant. In some yards where there are no spare stables it may only be possible to wash boxes when the horses have been turned away for the summer. Usually this is also the time to spring-clean, paint the stables and carry out any maintenance.

The muck heap is a harbour for germs and vermin and should therefore be sited at a suitable distance from the stable yard. A tidy muck heap is easier to maintain and quicker to rot if it is trodden down but an untidy muck heap will blow all over the yard.

Tack rooms and feed sheds should be kept clean in the interests of hygiene and tidy for the sake of convenience and easier management. An organised yard will have a place for everything and everything in its place, with safety and hygiene a priority.

Fire precautions and fire drill

One of the golden rules in the stable yard must be NO SMOKING, not even in the tack room. 'No smoking' signs should be placed strategically throughout the yard to ensure that no one is in any doubt, particularly in livery yards and riding schools which are open to the public. The muck heap and hay and straw barns are major fire risks, as are buildings with electrical points and anywhere where machinery is used or vehicles are parked.

It is advisable to invite the local fire officer to visit from time to time to ensure that everyone is familiar with the procedure in the event of fire. Likewise the fire equipment should be checked every six months and personnel must be aware of its location. Horses are terrified of fire and smoke and will panic in the face of it so it is essential that everyone knows the drill should fire break out.

Where staff accommodation is above or near the stables it is important that someone knows how many people are in the building at any time but especially at night.

The following list of fire precautions should become familiar to everyone in the stable yard:
Preventions:
Muck heap, straw and hay barn not too close to yard.
Sawdust bales not packed too close together.
Distance between stable blocks should be at least 8 ft (2.5 m).
Fire Extinguishers:
Blue = powder.
Green = foam.
Red = CO_2
Should be checked by fire officer every six months.

Should have valid date on them.

On every yard the fire hose should have sufficient power to reach every square inch of the yard.

Signs:

Displayed around the yard with exact procedure.

By law they should be white on lime green.

Fire procedure:

Should be carried out regularly for the benefit of new personnel.

Sand buckets should be distributed containing dry sand.

Drill:

Ring alarm.

Tackle fire if possible.

Turn horses into a safe field.

Telephone fire brigade.

Ensure someone meets the fire brigade on their arrival.

Fire extinguishers should be available anywhere where there is electricity.

Keep calm and don't panic.

If there is a choice between human and equine life, save the person.

Tools and equipment used in the stable yard

The most essential items in regular use are:

Fork. Three- or four-pronged fork either with a long or short handle which can be metal or wooden. Used for mucking out.

Pitch fork. This is two-pronged with a long handle and used mainly for bedding down straw.

Shovel. A large aluminium type is better than the smaller heavier metal shovel and is lighter to handle. Used for mucking out shavings, paper and peat bedding as well as for general stable use.

Yard broom. A hard-wearing broom with bristles made of either plastic or fibre. Plastic bristles usually last longer.

Soft broom. A household broom for use in the tack room, office and cloakroom.

Muck sheet. Custom-made muck sheets of plastic or canvas with corner loops for handles are available. Alternatively a large hessian sack cut open will do the job equally. Bran sacks are an ideal size but more difficult to find.

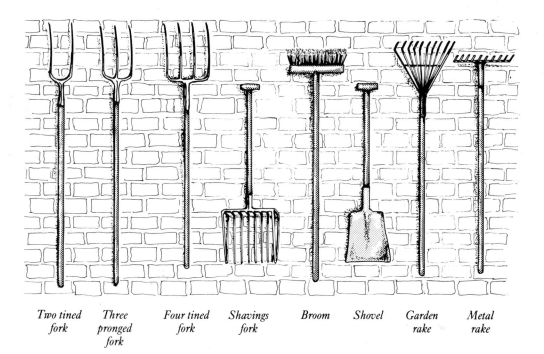

| Two tined fork | Three pronged fork | Four tined fork | Shavings fork | Broom | Shovel | Garden rake | Metal rake |

Plastic muck sheet.

Skep. Made of either wire mesh, plastic or rubber. The latter is most practical because it is safer should the horse tread on it and it is easy to clean. Choose one that can be carried with one hand.

Rubber skep.

Wheelbarrow. Shapes and sizes range from a two-wheeled garden type suitable for most yards to a farm barrow with a larger capacity. The choice will depend on the number of horses and staff and the type of bedding used.

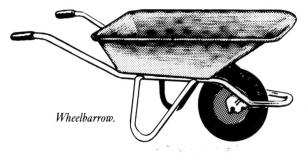

Wheelbarrow.

Rakes. The two most common rakes used in the stable yard are the light spring rake and the iron rake. The former should be used with care because although they are ideal for raking

gravel the prongs are designed for raking grass.

Hose-pipes. A heavy-duty rubber hose-pipe is more expensive than a plastic one but will last many times longer as the latter do split and are less flexible, should the horse tread on it. A pressure hose is ideal for washing down the stables but on no account should it be used on a horse. Hoses should be stored either on a wheel or wound up and fastened together with a clip or rope and hung up.

Water tank. A large tank with a plug is very useful for washing and soaking buckets, rugs etc. A galvanised tank is ideal but a cheaper alternative is an old enamel bath.

Stable bucket.

All tools should be washed weekly with fectant or hot soda water to prevent a bu of grease, especially on the handles. W barrows will need servicing too. The should be blown up as necessary and the oiled. Damaged tools should not be use they are properly repaired and not tempora tied up with string or wire.

Transport

The largest and most expensive piece of equipment in the yard is the horse transport; a lorry or trailer. It is usually the duty of the groom to keep the vehicle clean, which will involve mucking out each and every time it is used and washing it down regularly. A pressure hose is particularly useful for this but care must be taken not to remove the paint by getting too close. It is important to wash down the floor of the vehicle regularly otherwise the horses' urine will eventually rot it.

Maintenance checks should be carried out thoroughly and very regularly, i.e. weekly, in the interests of the safety of the horses and the vehicle's roadworthiness. Particular attention should be paid to the floors, ramps, doors and partitions and any necessary repairs should be put in hand immediately, before they become a hazard. Running checks, on such items as lights, brakes, fuel, oil, water, tyres and indicators, and to the towing hitch and safety chain on the trailers), must be done before each journey. Regular oiling of all movable parts will make the vehicle easier to use, especially singlehanded. Easy accessibility should be considered in case an emergency arises and the horses have to be taken off the vehicle in a hurry. Ideally the vehicle should be garaged but if this is not possible check it frequently for ... in the roof, signs of rust, or rot in any ... n parts. Electrical equipment should be ... ned in proper working order and any repairs carried out promptly.

Jumping equipment

Most yards have some fence material which is used for schooling purposes, either in a field or manège. There are various inexpensive alternatives to proper show jumps, such as drums, tyres, straw bales, sacks of straw, wooden pallets and oil cans. All these are potentially dangerous if they are not kept in good, safe order. If they become rotten they should be thrown away. Jump wings, poles and cavalletti will need annual attention with the paint brush and repairs carried out as necessary. Cups and pegs should be kept in a safe place and never left on the ground when not in use since there is a danger of a horse treading on them. A piece of wire or string attaching them to the wing will ensure they do not get lost in the grass. Ideally jumps should be stored indoors when not in use which will help them last longer. Alternatively they should be stacked tidily in a corner and if possible off the ground. All show jumps are expensive to buy and should therefore be looked after properly and not left to rot in a field and thereby become unsafe to use.

49

CHAPTER FOUR

STABLES AND YARD: DESIGN AND CONSTRUCTION

For practical purposes it has become convenient to stable the horse while he is being used. It is therefore necessary to have an understanding of the requirements in terms of the size and shape of room he needs. There are many materials from which to choose for building stables to suit each owner's particular situation. Usually the owner's available funds will influence choice and design. It may be necessary to budget for a building project which involves a number of stables. Whatever the case the owner should understand the principles of stabling horses, which is what this chapter explains.

Reasons for stabling a horse

Protection. A horse in a state of fitness does not carry protective fat. A horse that is clipped will have no protective coat and the action of grooming a horse removes many of its natural oils. Stabling will give protection against the cold, the rain and the wind in winter, and the sun and the flies in summer.

Convenience of management. The horse is at hand for regular attention and feeding, and can be kept clean and dry easily. The intake of food and water can be monitored and regulated as required. A confined area is desirable

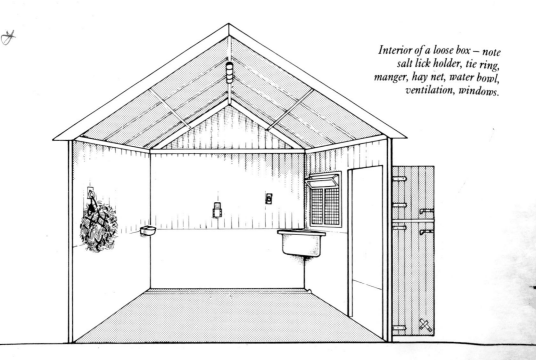

Interior of a loose box – note salt lick holder, tie ring, manger, hay net, water bowl, ventilation, windows.

Automatic water bowl.

Basic design requirements

The following measurements should be regarded as the minimum requirements for the housing of horses and ponies in a loose box:
Pony (under 14 hh): 10 ft × 10 ft (3 m × 3 m).
Large pony or small horse (e.g. 14–15 hh): 10 ft × 12 ft (3 m × 3.65 m).
Horse: 12 ft × 14 ft (3.65 m × 4.25 m).
Mare and foal: 16 ft × 16 ft (4.85 m × 4.85 m).
Individual stalls should be 6 ft (1.8 m) wide, 9 ft (2.7 m) deep with a 6 ft (1.8 m) passage behind.

Doorways must be a minimum of 7 ft (2.1 m) but preferably 8 ft (2.45 m) and 4 ft 6 ins (1.4 m) wide with outward-opening doors. Bottom half-doors should be 4 ft 6 ins (1.4 m) high.

The floor must not be slippery although it must be impervious to water, hard wearing and sloped or grooved for drainage.

The drains should be open-grated with a silt trap outside the box itself with a drainage channel through the partition.

The walls should be impact-proof up to 4 ft (1.2 m) high and have a smooth surface free from ribs and protrusions which may injure the horse and also invite chewing. They should be 10 ft (3 m) to the eaves.

The ceiling must not be lower than 10 ft (3 m), it should allow foul air to escape and preferably be of fire-retardant material.

for such things as veterinary attention and shoeing and is essential in cases of illness or where isolation is required. A stable can also act as a restraint, such as for an unruly horse or for a foal at weaning.

Security. This is an important aspect especially in the case of valuable horses. Stables with locked doors and a locked access route offer greater security than an open field. There is a danger, however, that over-emphasis on security could prevent horses being released quickly in the event of a fire.

Conservation of pasture. A field that has to support livestock over the winter will suffer damage to the sward. Excessive cutting up of the ground, known as 'poaching', will kill off many of the plants of grass and reduce the recovery rate of others resulting in a very late and unproductive crop of grass the following spring.

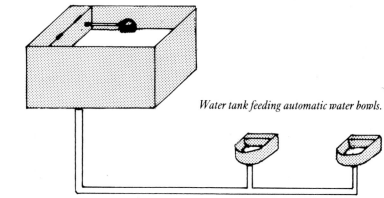

Water tank feeding automatic water bowls.

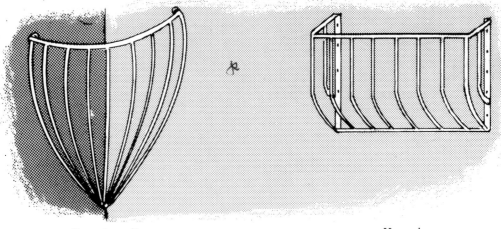

Corner hay rack.   Hay rack.

Corner hay manger.   Hay manger.

The fittings, i.e. manger, hay racks, water bowls and tie-rings, should be good quality, reliable, accident-proof and very securely fixed to the walls.

Lighting. Adequate lighting is essential and two lights per box are preferred to one. These must be of a tamper-proof design. Under no circumstances should power points or light switches be sited inside a loose box or be accessible to a horse over a stable door. Use industrial sockets and switches rather than domestic ones. Natural light from doors and windows will be enhanced by a window at the

Covered light bulb.

Protected light switch.

52

back and front of the box provided that the open top door does not obscure the front window. Grilles should be fitted to the inside of all windows to prevent the horse from touching the glass.

Ventilation is essential to good health but draughts are detrimental. In cold weather the body warmth of the horse should be maintained by the use of rugs rather than by closing all the doors and windows. The top door need only be closed against driving rain or snow provided there is adequate window ventilation also. The exception to this may be when a new or excitable horse is in danger of jumping out. Sheringham or 'hopper-type' open windows are preferred as they direct incoming cold air upwards to mix with the warm air. A through-draught is prevented by having the open window on the same side of the box as the open top door. Only open windows in opposite walls when a through-draught is required to reduce the temperature in the heat of summer. Some form of foul air exit is required and is best provided by a ventilated ridge, in the case of a pitched roof, or air vents high in the walls where there is a sloping roof.

Safety. One of the most dreaded occurrences in the stable yard is the outbreak of fire. Never let any of the aforementioned principles of stabling override the need for quick and practical means of dealing with such a disaster. Consider the location of flammable materials like straw and hay in relation to the animals. Avoid building long internal passages with only one exit door. Provide extinguishers, sand buckets etc. Discourage smoking by posting 'No Smoking' signs. Devise a fire drill and display it, showing the location of fire-fighting equipment and the telephone. Make it clear that the priorities are:

1. Save human life.
2. Save horses' lives.
3. Save property.

Detail your procedure, i.e. (a) get people and horses out; (b) telephone the fire brigade. This should be automatic for resident personnel but fires are often discovered and tackled by strangers who are unfamiliar with the layout of the buildings.

The stable block

This will consist of some or all of the following:

1. Loose boxes.
2. Stalls.
3. Feed room.
4. Hay store.
5. Bedding store.
6. Feed store – concentrates.
7. Washing room.
8. Drying room.
9. Tack room – ideally one for dirty tack and one for clean.
10. Utility box.
11. Manure bunker.
12. Office.
13. Lavatory.
14. Locker/changing room.
15. Staff lounge.
16. Store room – for mower etc.
17. Garage – horse box, Land-Rover etc.
18. Wheelbarrow park.
19. Fire control equipment, i.e. fire extinguishers, buckets of sand, fire alarm, fire drill notice, 'No Smoking' signs.
20. Yard clock.
21. Separate isolation box.
22. Specialist unit, e.g. stallion box, foaling box.

Construction considerations

When planning the erection of a completely new stable complex the following points should be considered:

The site

(a) Ensure that the road access is sufficient for large vehicles, i.e. feed merchants' lorries, horse boxes, and tractors with trailers. Not only must they be able to enter but also they should be able to turn round again to get out.

(b) Ensure that there is foot access to grazing areas for horses and that this is not too narrow or restricted.

(c) Allow room for any possible future extensions.

(d) Consider the availability of electricity and

water. Provision of electricity can be very costly if a supply is not already close at hand. Water must be available from a piped supply at a minimum of 18 ins (45 cm) deep to avoid freezing and may need to be metered if using mains water.

(e) Adequate drainage is probably the most important single aspect. Even the most luxuriously equipped stable block can be ruined by bad or inefficient drainage. The soil type will determine the effectiveness of your drainage system. A heavy soil may necessitate the drainage being piped to a carrier ditch whereas in a lighter, rockier type of soil an efficient soakaway may be effective, but make sure it is well away from the buildings and do not underestimate your minimum drainage requirements, particularly in the light of possible future expansion.

(f) Planning permission will be required in most cases and the above points should be accounted for when applying. In addition you may have to comply with tree preservation orders or any other mitigating conditions which apply locally. Trees too close to the stables will result in leaves blocking rainwater gutters in the autumn and could be considered a hazard if they are near enough to fall upon the stables if blown down in a gale.

Aspect and design

(a) The most ideal aspect is for the stable doors to face south or south-east in order to benefit from maximum use of sunlight and protection from northern winds. This is not difficult if the boxes are to be in a straight line but many yards have an L-shaped configuration or three or even four sides of a square. The chosen shape must fit the available site and will also determine any possible expansion.

(b) When erecting stables in a barn system, which is a very practical system to work with, the aspect is not so vital as long as the main doors do not open to the east or north winds, as in practice main doors tend to remain open for long periods during the day. With this system, ventilation is the prime environmental consideration. A number of horses kept collectively under one roof are prone to a rapid spread of respiratory infections in a badly ventilated barn. Good ventilation with adequate temperature control is not easy to achieve in winter time.

Foundations

When assembling pre-formed wooden boxes the only requirement will be a concrete 'pad'. This should be laid on plenty of hard-core, granite or limestone as opposed to any softer material which may break down in the course of time. The suppliers of the building will give recommendations as to the size and thickness of the pad but they will probably stipulate that it has to be exactly level which is not very helpful when it comes to good drainage. However, this can be improved by ensuring that the front 'apron' outside the boxes has an adequate slope and this is usually formed after the boxes are erected. Probably better are the boxes that are made to sit on a low wall of concrete blocks. This allows the blocks to be laid on trench footings and the floor of each box to be concreted individually and grooved to a drainage point. The same applies if the stables are to be built entirely of concrete blocks or bricks except that the trench footings must be more substantial.

Construction materials

Walls

(a) Stone. Probably the most aesthetic but definitely the most expensive to erect. Warm in winter, cool in summer. Vermin-proof and horse-proof.

(b) Concrete blocks. Cheap to buy and erect. They can be painted to enhance their appearance and are maintenance-free. It is advisable to fill the holes to prevent the blocks from cracking if kicked by horses and to deter vermin from nesting in the stable. Concrete blocks are particularly cold in winter but this can be overcome by an interior lining of plywood.

(c) Brick. More expensive than concrete blocks to buy and erect but infinitely more aesthetically pleasing. In suburban areas planning permission may require the use of a certain colour or type of brick to be used in environmentally sensitive areas, but this may only be necessary for the front aspect. It is not uncommon for a stable block to have an attractive brick frontage and the back and side walls to be of concrete blocks or timber. A cavity brick wall can be insulated and be much warmer than any other medium.

(d) Timber. Relatively expensive, easy to erect and repair but also easily damaged. It is subject to being eaten by horses and vermin and needs regular treatment to remain waterproof. Its main disadvantage is that it is a tremendous fire hazard. An asbestos roof helps to reduce this slightly but timber stables with timber and felted roofs are to be avoided.

Roofs

(a) Asbestos. Relatively expensive, easy to erect but very brittle and easily broken, yet maintenance-free. When used in a livestock building it is advisable to insulate the underside to avoid condensation dripping onto animals below. This can be done cheaply with wire netting and straw as well as other proprietary materials.

(b) Corrugated iron. Cheap yet effective. Being of lighter weight will not require such heavy support. Will need painting if not of the plastic-coated type. Can be noisy and frightening to horses in heavy rain or during a hailstorm and is liable to create condensation. Insulation, as with asbestos, will be helpful on both these counts.

(c) Perspex sheets. These have the advantage of being lightweight but unless they are extremely well secured may be blown away in a high wind. They are expensive and come in a variety of colours and can therefore create an attractive building. They are not liable to condensation as much as asbestos or iron and clear transparent sheets can be used within roofs of other materials to provide more natural light inside the building.

(d) Slates or tiles. The most resilient and attractive roofing material, maintenance-free yet expensive, requiring much more substantial roof support by virtue of the extra weight. When used with a felt underlay, totally weatherproof and warm.

(e) Timber and felt. Cheap but unsuitable as felt is not very durable. It will allow water in when holes appear and timber can rot unless the felt is repaired or replaced. Carries a very high fire risk.

Floors

(a) Concrete. Probably the most common in current use. Needs a good base of hard-core and must itself be a minimum of 4 ins (10 cm) thick. Can be very cold and damp for horses to lie on but this can be greatly improved by the use of a damp-proof membrane, i.e. a sheet of polythene laid with the concrete. Stable floors can be sloped and/or grooved in a herringbone fashion to assist drainage.

(b) Vitrified bricks (blue bricks). These are comparatively expensive and need to be very expertly laid in a bed of sharp sand to be successful. They are warm to lie on and do not hold water but their durability can be suspect depending on their source.

(c) Chalk, sand or soil. These have the advantages of being softer and warmer to lie on but are also dirtier, less hygienic, dusty and/or muddy often at the same time in different parts of the box. Cheap.

(d) Timber. Undesirable from the point of view of durability, splintering, rotting and providing habitat for vermin.

(e) Cobblestones. Seldom used nowadays. Relatively expensive to buy and lay. Can become slippery when wet and rather uncomfortable.

Isolation box

The purpose of an isolation box is to confine the sick and infectious animal until such time as he is no longer a health hazard. The box (or boxes depending on the number of stables available in the yard) should be as far away from the other boxes as possible and practical. The ideal size for an isolation box is 16 ft × 16 ft (4.85 × 4.85 mm) to allow for any veterinary treatment; anything smaller may hinder the work of the vet and restrict the horse if he has to be confined for long periods of time. The fixtures and fittings will need to be the same as for any other box. The ventilation is critical because the sick horse, more than any other, should not be subjected to draughts or stagnant air. The type of material used in the construction should be the most practical to allow for washing the box down thoroughly and disinfecting it after each confinement. Steam-cleaning is ideal and the relevant equipment can be hired, although a pressure hose will do an excellent job too. It is particularly important that the sick horse is kept as warm as possible without him having to wear too many rugs. His bedding should be generous particularly if he has a leg injury, in which case he will be less restricted with shavings rather than straw. If one can afford it the walls and floor of the isolation box can be covered with rubber to safeguard the animal against injury. Again if money is no object the sick box could have a small paddock attached to enable the horse to exercise himself alone once his condition allows him.

The isolation box should be used for any horse who has an infectious disease, such as strangles, a cough which is accompanied with a runny nose, and skin diseases such as ringworm and pox. It is important that only one person attends the infectious horse whilst he is in confinement to avoid the danger of an infection spreading to the rest of the horses. The attendant should be sure to disinfect him/herself after handling the patient and keep all kit such as grooming kit and any other equipment separate to avoid contamination. Like the stable all of this equipment should be washed and disinfected thoroughly after each confinement.

The muck heap and disposal of manure

Every yard will have to decide on the best siting of the muck heap depending on the type or types of bedding being used and the method of disposal. For the purposes of convenience it will need to be as near to the main stable block as is practicable so that time is not wasted to-ing and fro-ing, particularly during mucking out. Preferably it should be out of sight of the house and front of the yard as vermin and flies are always attracted to it apart from it being unsightly. If the manure is to be removed at a later date it will be necessary to allow access either by a tractor and trailer or lorry. Both will need room to manoeuvre their loading machinery, such as a front loader or hydraulic grab. The problems of disposing of manure are worse in suburban areas where neighbours are often understandably offended by bonfires of shavings or sawdust, which do take a long time to burn out. Peat, on the other hand, is more easily used on the land and can often be sold to gardeners. Straw, like shavings, can be burnt if there is somewhere suitable to site a fire. Alternatively it can be sold to mushroom growers who will collect. If you are farming you may be able to deposit it onto the land but it will need to be well rotted down otherwise it will blow everywhere and be of no value as manure.

The size of a muck pit will depend on the number of horses kept. If funds allow, a construction of, say, breeze blocks on three sides with a concrete base is ideal. If the floor is to be left on earth it must be lower than the surrounding land to contain the effluent. The important thing to remember in construction is to slope a concrete floor in such a way as to direct the effluent away from buildings and preferably on to land as it will eventually rot materials such as wood and concrete. Where possible a hedge planted on three sides around the muck pit will camouflage it without hindering access by machinery.

The maintenance of the muck heap reflects the standard of stable management. It should be tidied and trampled down regularly to encourage the manure to rot and to keep it compact. This job should be someone's responsibility after mucking out every day. A muck heap can very quickly spread and look untidy if neglected.

THE TACK ROOM

Security

For the purposes of security as well as convenience it is advisable to site the tack room as close as possible to the stables and main complex of buildings and, if practical, near the house too. Tack thefts have become all too commonplace today so it is prudent to have all the tack marked with your post code. Do not use your name on tack because it cannot be identified this way. By marking tack there is more chance of it returning to the rightful owner if it is ever recovered. Apart from fitting an alarm system, which is not financially within every stable owner's reach, a good guard dog is the next best defence against intruders. Iron bars or grilles should be fitted and the tack-room door kept locked at all times when there is no one in the yard. This includes the times when horses are out being exercised because thefts take place as much in the daytime as at night. The local police will be happy to visit and advise you on the best form of security for your yard. If you are away at competitions and the yard unattended it would be wise to notify your local police. Wheelbarrows and trucks are often used by burglars to transport tack from the tack room to their vehicle so it makes sense to keep these locked away when not in use.

It is helpful to maintain an inventory of all saddlery and equipment, perhaps in the stable diary with a duplicate kept in a safe place at home.

Fixtures and fittings

One of the most essential items in the tack room is the saddle horse. Saddles can then be conveniently and safely cleaned, without fear of damage, before storing them on a saddle rack. Saddle racks are best fitted to the wall above head height to save space. An adjustable hook suspended from the ceiling is necessary for hanging tack to be cleaned and this will need to be sited in a convenient place for working. Bridle-hooks for hanging bridles, headcollars, martingales etc. should be fitted to walls at a height which will allow reins and other long pieces of leather to hang without touching the floor. Cupboards and shelves are essential for storing sundry items such as boots, bandages, rugs and rollers. A veterinary cupboard will also be needed to keep medicines in a clean, dry and hygienic state.

The floor is easier to keep clean if it is laid with linoleum or tiled. Lighting will need to be good; it is hopeless trying to work with poor lighting. The room temperature and atmosphere can have a detrimental affect on any leatherwork stored therein: dry air will make the leather go hard; dampness will cause mould to form. Some form of heating will therefore be necessary; either a radiator, stove, bar heater or other suitable appliance which would be safe if left unattended. It is useful to have some form of electric drier for rugs and bandages, boots and numnahs and so on, also a clothes horse or rack for airing damp items. The most convenient as a space saver is the type which is suspended from the ceiling by a pulley and can be raised and lowered as required. A large sink, preferably with hot and cold running water, will add to convenience when tack cleaning etc. Power points are also useful for boiling kettles or re-charging battery clippers. Large chests or trunks are best for storing rugs, which can take up a large amount of room. Remember, rugs should not be stored without moth balls because they can be badly damaged in no time at all; the same applies to bandages and numnahs.

Somewhere in the yard, preferably in the tack room, a telephone should be available, not least of all for emergencies. A fire extinguisher must be kept in a well-sited place for easy access. Cleanliness and tidiness must be a priority in the tack rooom as much as anywhere else. All tack and equipment is expensive and should be kept in best condition for economic as well as safety reasons.

The tack room is the best place to keep the stable diary, described in Chapter 3.

CHAPTER SIX

THE FEED SHED

General considerations

The location of the feed shed should be as central as possible to the main stable complex. In the case of a barn-type yard it could be in the middle of the line of boxes. A lot of time is wasted going to and from the feed shed if it is not conveniently sited.

Materials for the feed-shed construction will be either bricks or wood. A damp-proof course in the floor is recommended, to keep the shed dry. Accessibility for deliveries is important so there should be enough room outside to allow lorries to park and turn around.

The size of the feed shed will depend on the number of horses kept, but for practical purposes it will need to store at least one week's feed. An average size would be 14 ft × 16 ft (4.25 m × 4.85 m) for a yard of up to ten horses. A wide door affords better access for carrying sacks; an 'up and over' garage door is ideal.

A water tap, either inside or nearby outside, is essential, but where funds are available it is both convenient and labour-saving to supply the feed shed with hot and cold running water and a large sink. The feed-shed area should be kept clean and tidy with utensils and feed receptacles regularly washed.

Good lighting is important and a large fluorescent strip is ideal. If the roof is fitted with Perspex sheets it will help make the shed light and save on electricity.

Vermin control and cleanliness

As far as possible the feed shed should be vermin proof, although this is difficult to achieve because vermin can chew their way in through woodwork. A pest-control programme should be employed because vermin can cause many pounds' worth of damage and are a constant health hazard. One of the most common diseases which can be transmitted to horses and dogs via the rat is leptospirosis. It is therefore important to ensure that the feed shed is swept out after every feed time and room is left around the bins and sacks for the stable cat or dog to seek out its quarry. Waste collected in a bin or sack should be disposed of once or twice a week. Stale or left-over feed can be deposited on the muck heap but should never be left in the feed shed. It is a good idea to hang a brush and shovel in the feed shed to encourage cleanliness.

Fixtures and fittings

Provision should be made for storing feed buckets and spare mangers and a cupboard provided for keeping feed additives and supplements cool and dry.

A feed chart is an essential feature of the feed shed. The most practical type is a blackboard on which every horse's diet, including exact quantities and time of feed, can be written. It is vitally important to keep the feed chart up to date and ensure that any changes, however small, are noted, even temporary alterations such as when a horse is rested and put on a laxative diet. Failure to do so can have disastrous consequences where the same person does not make up the feeds.

An electric socket is useful for boiling a kettle for a mash or for an electric boiler, commonly used for cooking barley, oats and linseed. A fire extinguisher must be kept within easy access of the feed shed, preferably inside it.

Some larger establishments, especially farms, have their own oat crusher which enables oats or barley to be rolled at regular intervals. This is particularly useful because rolled oats and barley will have lost some of their nutritional value if they are more than a week old when fed.

Storage

Storage bins, made of wood, plastic or galvanised steel, must be vermin-proof. They should be cleaned out regularly to prevent damp and mould forming, thereby contaminating the feed. Feedstuffs do not keep indefinitely and it is sensible to order small quantities at frequent intervals to avoid wastage. Bins with compartments which will hold two or three different types of feed are practical and space-saving. Feeds such as bran, sugar-beet and nuts which are sold in paper sacks must be stored off the floor, e.g. on wooden pallets, to safeguard against damp and mould.

Nuts should be kept cool and dry, preferably in metal bins which have been raised off the ground. Old nuts should be cleaned from the bottom of the bin before refilling with fresh nuts. Even if you buy your food in bulk and your bins are large, only put into the feed bins enough feed for, say, a week and store the rest unopened in the sacks in which it is delivered. Never put fresh food on top of old otherwise the old will eventually become stale and may go mouldy making it both unpalatable and dangerous to feed. Nuts, particularly, will soon develop a foul smell when they go stale and should not be fed under any circumstances as they may cause colic. Many horses will not eat stale nuts but this must not be used as a measure of their staleness.

THE HOUSING OF BEDDING MATERIALS AND FODDER

The storage area considered

Facilities must be made available for storing all bedding and fodder under cover, and preferably protected on at least three sides from the weather. A Dutch barn is suitable for housing straw, hay or shavings bought in large quantities. Any such barn should be situated away from the stables because of the fire risk yet it must be practical from a working point of view. It will need to be accessed with a tractor and trailer or lorry not only from the ground but also avoiding overhead cables. Dutch barn bays are a regular 15 ft (4.5 m) wide and can be added to according to the amount of bedding and fodder to be kept. Lean-to's which are built on to existing buildings are very useful providing they are not attached to a stable block, because of the fire hazard. The 'No Smoking' regulation should, of course, extend from the stable yard to the barn area as this has the highest risk.

In barn-type stabling fodder and bedding can be stored integrally either in a spare area or on a false floor if there is enough space in the roof to fit one. In large establishments where room allows, it is convenient to bring into the barn a week's supply of bedding and fodder at a time from outside buildings. Indoor schools can also be used to store bedding at one end and can prove to be an asset at harvest when, for the sake of storage, it is more economical to buy hay and straw off the field.

Hay and straw

Hay and straw are generally more expensive in the spring and therefore it is worth considering making storage space available in order to be able to buy larger quantities when they are cheaper.

Bales should be taken from the stack layer by layer and not just from the front. Not only does this help to save wastage but it also makes for safer handling of the bales. Any loose hay or straw should be used up immediately and not left to accumulate and be wasted. It will be trodden on, become dusty and blow around the yard.

Loose string is potentially hazardous so bales which are opened in the barn should have the string removed and tied in a knot before depositing it in a sack or bag provided for this purpose.

Stacking hay and straw outside under a plastic sheet is not recommended because the weather will cause considerable wastage. The bottom layer of any hay which has been stored in outbuildings is usually unfit to feed to horses but can be given to cattle or sheep. For this reason it is good practice to first put down a layer of straw underneath the hay. The bottom layer should be placed on its side to prevent the string from rotting.

Shavings

Shavings are available either loose or in plastic bales and hessian sacks. Plastic bales can be safely stored outdoors, providing they are not punctured or damaged by vermin. Once the plastic is pierced the bale must be used immediately before the shavings become wet. Rat bait can be placed between bales during stacking and inspected regularly. Hessian sacks will need a dry storage area, perhaps under a Dutch barn if building space is not available. Again they will need to be protected from vermin by putting poison down and checking it regularly. Once the sacks are empty they should be tied in regular bundles. Plastic bags should be burnt on an incinerator or open fire.

Routine maintenance

All barns and storage areas should be cleaned out thoroughly at the end of each season to prevent mould and rot. There will inevitably be a build-up of spillage and waste which should be taken clear of the buildings and burnt or removed to the muck heap.

Emphasis must be placed on the control of vermin in storage areas and around the stable generally because the damage they can cause can be considerable, apart from their being a health hazard. Prevention is better than cure and a pest-control service is a worthwhile investment as it will keep the situation in check all the year round.

It is a requirement of the Health and Safety Order that ladders are maintained in sound working order. Any broken rungs must be mended promptly and the ladders not used until they are once again in safe repair.

DUTIES RELATING TO MUCKING OUT

How to use stable tools and equipment safely

Whatever stable chore is being carried out, the safety aspect must always be considered both from the horse's point of view as well as the human's. Using tools of any sort can be hazardous if the user is careless or neglectful. The tools used for mucking out stables can cause serious harm if left accessible to the horse, and no risks should be taken such as leaving them in the box with the horse. The following list includes some of the golden rules for using tools and equipment in the stable yard:

1. Never leave any tools or equipment, albeit for a moment, where the horse can reach them.
2. Always store tools and equipment in a safe place which is both handy and out of the way of passing horses or people.
3. Never use any tools or equipment which are in need of repair.
4. Always be sure that repairs to any tools or equipment are of a permanent nature. Temporary repairs are a safety risk. A piece of baling string wrapped around a broken handle will not do.
5. Never leave a wheelbarrow where a horse can reach it, for example in the doorway during mucking out if the horse is loose in the box.
6. During mucking out always be sure that tools which are not in use are left outside the stable door.
7. Whenever a skep is in use do not let the horse walk on it and possibly frighten or injure himself.
8. Never use a fork near to the horse, always move him out of the way to allow you to work safely.
9. Any string or bags should be cleared away to a safe receptacle and never left on the yard. String should be tied into a knot.
10. All combustibles should be incinerated regularly and other waste deposited in the dustbin or at a recognised waste tip.

These are the tools and equipment most commonly used for the chores which relate to mucking out:

Forks: four-tine – for mucking out straw, peat, bracken and newspaper.

   three-tine – suitable for mucking out and bedding down straw, peat, bracken and newspaper

   two-tine – for bedding down straw and bracken.

   shavings fork (otherwise known as a potato fork) – used for mucking out and bedding down shavings, sawdust and newspaper

Shovel – necessary for handling shavings and sawdust and on the yard in general.

Brush – a yard broom is essential for all sweeping chores on the yard.

Rake – a garden or lawn rake is useful where there is a gravel yard.

Wheelbarrow – necessary for mucking out and other yard chores.

Muck sack – an alternative to the wheelbarrow.

Skep – a small container which is used for removing droppings during the day. Can be made of rubber, polythene, plastic, wire, wood or other material which is durable yet light and easy to carry in one hand.

Haynet – where fixed mangers are not in use the haynet is an alternative means of feeding hay off the ground. It can be made of string, tarred rope or polythene twine.

Types of bedding material

Straw

There are three types of straw which can be used for bedding horses, listed here in order of practicality.

Wheat straw is the first choice. It is best suited for horses because it is harder and shorter than other types. As a rule the horse is less likely to eat wheat straw than he is to eat others.

Barley straw is less preferable because it is palatable, although less digestible, and horses are always tempted to eat it. It can very easily ball up in the horse's gut and cause colic. Fatalities have been recorded due to balled-up barley straw in the gut. The husk of barley can also be harmful if it gets into the horse's eyes.

Oat straw is again palatable to the horse and therefore not desirable. The straw length is longer than other types and more difficult to handle. Bales are invariably larger and therefore heavier.

Whichever type of straw is used the bale should be shaken up thoroughly and a thick bed with high banks around all the walls should be built. A straw bed is easily displaced once the horse turns about and rolls so it is important that there is sufficient straw to give the base a good foundation and to guard against the horse injuring himself. It is less absorbent than other types of bedding and relatively more is needed to make a good deep bed every day. Wastage is high compared to shavings, sawdust and peat. The main disadvantage of any type of straw is the dust content, even present in what may appear to be 'clean' straw. This is harmful to the animals' respiration and one of the reasons why many horse owners, in particular racehorse owners, have switched to shavings. It is, however, dry and comfortable for the horse, easily obtainable, less harmful to the hoof, as well as being easy to handle and muck out. It will burn if necessary or can be left to rot down and sold to mushroom growers, farmers etc.

Shavings

Wood shavings from saw mills are becoming increasingly popular for stabled horses because they create less dust than straw and owners are now more aware of the danger to the horse's health caused by a dusty atmosphere. Respiratory ailments often result from the horse being subjected to a confined area which is continually dusty. Consequently the horse's fitness is endangered and this, if not given due consideration, can effect long-term damage. One of the common diseases which has been identified in recent years is COPD (Chronic Obstructive Pulmonary Disease).

Although shavings have this advantage over straw they are more difficult to handle. They are usually packed in large polythene bags which are awkward for one person to lift and carry. Some saw mills like you to bring your own sacks and bag your own shavings, which is more economical. There are many companies now who distribute shavings in bulk and if you have somewhere dry to store them large quantities can be delivered loose.

Disposal of wood-shavings manure is not terribly efficient because one often has to rely on burning it. Ideally it should be removed from the stable area to a convenient site where it will not offend neighbours whilst it is burning. Shavings manure is useful for laying over grass to provide a surface for lungeing horses on when the ground is hard. It can also be applied to some outdoor arenas to improve the going.

Many yards use wood shavings for deep-litter beds because shavings are better than straw for soaking up moisture. However regularly this type of bed is mucked out a deep foundation of 6 ins (15 cm) minimum as a shavings bed is essential to encourage the horse to lie down and to safeguard against injuries. If a horse is prone to eating a straw bed, shavings are a useful alternative.

Sawdust

This is much finer than shavings and consequently more dusty. It is also more prone to pack in the horse's feet and to block up drains as well as being slower to burn. With the increased availability of shavings, sawdust is less popular today. A bed should be no less than 6 ins (15 cm) deep in the middle with high banks around the wall. Compared with straw, it is labour saving from the point of view of mucking out. Sawdust, like shavings, should be checked thoroughly for any foreign objects which may have passed through the saw mill and which could be harmful to the horse. Whilst it is less practical to handle than shavings it is also less expensive and for this reason may suit horse owners with a limited budget. It can be useful for soaking up spilt oil etc. and for use in the horse box instead of straw.

Newspaper

In recent years the use of waste newspaper as a form of bedding has afforded the horse a dust-free bed which, given sufficient covering, is no less attractive for him to lie upon and has the capacity to absorb moisture equally as well as, if not greater than, traditional alternatives. Its use is not as widespread as it might be purely because of its availability but owners who have horses with respiratory problems are keen to use it as the safest dust-free bedding currently on offer. Its main drawback at this stage has proved to be its disposal, which relies on convenient incineration. It can be awkward to handle and will blow away easily. For this reason it may be prudent to use a muck sack for mucking out.

Peat Moss

The use of peat moss in bedding down stabled horses is less common today than it was before shavings became so popular. Its chief advantage must be its highly porous properties for it will absorb both liquid and atmospheric gases, such as the ammonia emitted from urine. Its dark appearance gives the box a dull look. The less observant groom who is not as thorough as he might be which will soon cause a build-up of fouled bedding. Once established it offers the horse a comfortable bed but again it must be deep to withstand the horse turning and digging it up. It requires careful management with regard to the horse's hooves because it balls up easily in the foot and can absorb the natural moisture in the hoof. It is also blamed for rotting the frog of the hoof if neglected. If the hoof is not carefully picked out just before the horse is taken out of his box, the hooves will empty peat on the yard and create more work. This, however, can be said of all bedding but peat is particularly bad in this respect. The blame here though is on the groom and not the bedding.

Peat moss bedding is dusty and seems to encourage the horse to roll more than other types of bedding. For this reason horses kept on peat moss bedding are generally more difficult to keep clean and to maintain a bloom to their coat. It discourages the horse from eating its bedding and a good groom will find it quick and easy to muck out. Droppings should be picked up even more regularly than usual to avoid the horse treading them into his bed where they will be easily lost.

Peat moss is usually sold in plastic-wrapped bales or bags which can be stored outdoors. However, if they begin to smell as a result of dampness penetrating them they should be opened out to dry and stored indoors.

How to handle and carry weights safely

The nature of stable work requires the lifting and carrying of heavy and often awkward weights such as bales and sacks, particularly at mucking-out times. In order to be able to perform these chores with the minimum of effort and maximum safety it is helpful to know just how best these jobs can be dealt with. Many accidents are the result of careless short cuts and injuries can occur, especially to a person's back.

Whenever possible bales and sacks should be transported on a wheelbarrow but often, for the sake of convenience and speed, one bale or sack is carried on the shoulder. To minimise the danger of injury any weight should be picked up from the ground by standing in front

How to lift a weight correctly – note bended knees.

How to carry a bale using a pitch fork.

of the weight to be lifted and bending the knees. Whenever possible the weight should be lifted straight up, high enough to rest on something so the person can turn before taking hold again and carrying it on his shoulder. Always avoid lifting a weight from ground level, swinging it onto your shoulder and at the same time twisting your back. This is very often the cause of injury to the lower back. Weights should only be carried on the back for a short distance because of the one-sided weight distribution which again is harmful to the back. Small sacks which can be lifted high and held against the chest, can generally be carried comfortably but if the person has to lean back to support the weight it is too heavy to carry in this way and should be put on a barrow.

A wheelbarrow should be loaded with the weight towards the front, over the wheels, and again the item should be lifted from a stooped position. Saddles can be awkward to carry

especially if, as often as not, you are trying to carry other things at the same time. It is best to carry them in front of you rather than to the side. Try not to carry too much at once whilst loading or unloading the horse box. Avoid attempting to carry trunks and chests in and out of the horse box singlehanded. They are not only heavy but are awkward to manoeuvre. Whenever possible buckets, particularly those full of water, should be carried one in each hand for equal weight distribution. A full muck sack should first be lifted onto something, perhaps the manger, before turning and getting your weight underneath to carry it on the shoulder.

The principles of mucking out

Every horse keeper and groom should understand the importance of correctly mucking out a horse's stable each and every day, with no exceptions, if the horse is to be confined in sanitary conditions. His welfare relies on a hygienic environment which is maintained at all times if he is to remain healthy and in a thriving condition. Badly cleaned out and littered stables predispose a host of diseases and injuries as well as attracting vermin. A polluted atmosphere caused by the ammonia gases emitted from manure, is harmful to equines and humans. Ventilation plays an important part in the stable's hygiene and is considered in Chapter 4. To be sure that the stable is kept as fresh as possible whilst ensuring the horse's comfort, the manure should be removed from the box each morning, preferably at early morning stables as the first chore of the day after feeding, and the box left to air for as long as is practicable.

How to muck out efficiently

With the horse to one side of the box, first remove all the droppings with a muck or shavings fork without taking any more clean bedding than necessary. The clean bedding should then be moved to one side of the box tidily. Once you have exposed the soiled bedding it can be removed from the floor and stacked onto a wheelbarrow or muck sack. Each section of the box should be cleaned out in this way by putting aside the clean bedding and turning out that which is dirty. The floor can then be swept thoroughly and left to dry.

Slip-straw

If the horse is to stand in his box, slip-straw or shavings can be put down to safeguard against slipping. This is a thin layer which covers the exposed floor but does not prevent it from drying and airing. Slip-straw is only intended to be used as a temporary measure until such time as the box is bedded down fully.

How to load a wheelbarrow

If a wheelbarrow is loaded properly it will carry three times as much as one which is not. When mucking out, if the droppings are removed first these can be placed in the well of the barrow. The dirty straw can then be loaded first at the front over the wheels, and then evenly over the barrow, starting with the outsides and working inwards. It can be stacked quite high as long as it is level and packed down by treading on it or using the muck fork to pat it down. Try to avoid having straw hanging over the sides as this will probably fall off once the barrow is moving.

Use of a muck sack

Muck sacks used to be made by cutting open large hessian bran sacks but as more and more feed sacks are made of paper it is becoming increasingly difficult to find a sack which is large enough. It is, however, possible to buy from garden centres, a tarpaulin or plastic sheet with rope handles on the corners, which serves the purpose perfectly.

A muck sack is easy and convenient to use for mucking out by spreading it flat on the ground by the stable doorway and loading it with enough manure to carry comfortably to the muck heap. It is also tidier than a wheelbarrow for carrying straw, hay or shavings because they cannot blow about on the yard. Once the muck sack has been used for manure it should be left to dry out. A separate one should be used for carrying hay to avoid contamination from manurial odours.

Skepping out

To skep out is to remove the droppings from the loose box or stall without taking any bedding. By placing the skep at an angle the groom should be able to lift the straw underneath the droppings and cant the dung into the skep. The groom should use a skep whenever the horse is attended and at regular intervals throughout the day as a matter of habit. This helps to keep the horse's hooves free from dung and saves the bedding from becoming unnecessarily soiled. An efficient groom can skep out a straw bed without a fork, taking almost no straw at all. The full skep can then be emptied onto a wheelbarrow and taken to the muck heap.

Disposal of manure

The disposal of horse manure should be considered before you choose which type of bedding to use, according to your circum-

ABOVE: *How a muck heap should be maintained.*   BELOW: *How a muck heap should not be maintained.*

stances. Availability and economics will play their part but the manure must be disposed of in a practical way.

As discussed in Chapter 4, straw manure can be burnt, sold to mushroom growers as a form of fertiliser or, if you have the land and equipment, can be spread with a muck spreader on land which is to be ploughed. Shavings and sawdust are more difficult to dispose of if burning is not possible. Peat can be used on the garden as a fertiliser.

Sweeping the yard and ancillary buildings

This is a chore which should become part of the daily routine and any owner should insist on a thorough job being done especially after mucking out is completed. All the stable yard area should be swept and/or raked before breakfast, lunch and at tea-time and the sweepings carted to the muck heap in a wheelbarrow, skep or muck sack. Apart from the unaesthetic appearance, a yard littered with waste fodder, manure and bedding will quickly harbour vermin and become a health risk. This is particularly important in the feedshed area or anywhere where food and fodder is stored. Vermin can soon ruin large quantities of food causing many pounds' worth of damage.

Safety is another reason for maintaining the stable yard in a tidy order. Items such as baling twine, wire and paper can all become safety risks if they are left lying about. Sadly one often sees a horse shy at paper, or hears of animals cut by wire due entirely to human negligence.

Good stable management begins with the horse's welfare and a large part of that is down to basic housekeeping. It is far less effort to keep an establishment tidy on a day-to-day basis than to leave things to build up and involve a lot more time and work.

Disinfecting stables and drains

In order to effect a degree of sanitation where horses are kept it will be necessary to flush the drainage areas at least once a week with cold water to ensure they are flowing properly, before dousing with liberal amounts of disinfectant. This applies to any sinks as well as to the stable floors themselves. It may not be practical, from the point of view of drying, to wet the stable floor thoroughly where the box is used constantly, especially where absorbent bedding is used, i.e. shavings, peat and sawdust. It is necessary to remove these types of bedding altogether before washing down a stable floor and for this reason it is perhaps only feasible to do it once or twice a year. The strong stable odours emitted from the horse and its manure will be unpleasant enough even with correct ventilation and ample fresh air, so there can be no room for ignoring the need for regular disinfecting in the stable area.

A watering can with a rose attached is a handy way of dousing with a disinfectant/water solution.

Cleaning of water receptacles

Daily cleaning of all containers which are used for the horse's drinking water should be carried out at early morning stables, i.e. immediately following mucking out. Buckets will need wiping out with a sponge or cloth to ensure that any film or slime is removed. The same applies to automatic drinking bowls, which quickly build up a covering of green slime and harbour germs. In the case of fixed water bowls it will be necessary to empty the receptable first with a bucket or bowl. It has been written elsewhere that the drinking water should be changed for fresh water at least three times a day. Drinking bowls should be checked at regular intervals throughout the day to ensure they have not become fouled or blocked up with bedding or hay and that the water is flowing freely. The main disadvantages of automatic water bowls is that they can malfunction and deny the horse water. It is also impossible to monitor the horse's intake, which is sometimes an early symptom of ill-health. Unless the water pipes connected to bowls are well protected, freezing can occur during the winter months, which will mean that water buckets will have to be used instead.

Placing of water buckets in the stable

The safest place for a water bucket in the stable is at waist level, fitted to the wall by a custom-made holder or clip. There is always a danger that horses will kick over their water bucket and catch a foot in the handle if the bucket is placed on the floor. The bucket is also liable to catch droppings which will foul the water. If the bucket has to be set on the floor it should be right in the very corner with the handle flat down away from the horse. The corner furthest away from the manger and hay rack is preferred to discourage the horse from dunking food in the water. If a horse has a habit of dragging the bucket across the floor even when it is full, the handle could be clipped to a staple on the wall providing the horse cannot play with the clip.

Cleaning of mangers and disposal of left-over food

Mangers in regular use should be cleaned out at least daily, e.g. after mucking out each morning, using a sponge or cloth. If cleaning is ignored for a few days the remains of feed will collect, making it necessary to soak the manger before scrubbing it out. A dirty manger will quickly put a horse off his feed, just as we would be put off our food if it were continually served on a dirty plate.

If food is left from a previous feed this should be disposed of on the muck heap and preferably buried to discourage flies and vermin. Left-over feed should never be placed where animals could reach it. Stale feedstuffs, such as bran, sugar-beet and cooked foods, are particularly dangerous to the horse and can cause colic.

Tying up of haynets

A filled haynet must be tied up safely and securely (not less than 6 ft (1.8 m) from the ground) because if it hangs low it will be a hazard to the horse, especially when it is empty.

Once it is filled the tie string should be pulled as tight as possible to close the net. The tie string should then be passed through a ring on the wall and looped through a lower part of the net. After pulling it tight a quick-release knot should be made before turning the net round so that the knot is against the wall.

Any broken parts of the net should be repaired before a hole becomes large enough for hay to fall out and possibly be wasted.

Deep litter

It is entirely possible to keep horses on a deep-litter bed in the same way as cattle, but it is not ideal for the horses' feet. For the purpose of hygiene and care of the horse's feet, a deep-litter bed should have the droppings removed daily and fresh bedding added. Straw, shavings or peat can be used as bedding but if they are not maintained on a daily basis they will become unmanageable and very difficult to clean out. A deep-litter bed will need mucking out completely from time to time, which is hard work. It is, however, convenient as a labour-saving method where time is short or if a horse for some reason has to stand in for a while and needs the protection of a deep bed. It will require particular attention if the smell becomes offensive and a health risk, and vermin may be attracted to it, especially in winter time.

GROOMING

The principles of grooming

One of the most important aspects of stable management must be the essential duty of grooming working horses. No matter what the horse's role in life if he is employed he will require a daily grooming to maintain him in good condition. Before the horse's fitness comes good health, which can only be maintained by correct and thorough care, of which a large part is accreditable to good grooming. Sadly, in more recent times with the increased costs and restricted time which some owners have, the time spent on grooming is not as generous as it once used to be when stables employed one groom to two horses. It must be said, however, that in those days the groom was usually a man who often did not ride and therefore had more time to spend on stable duties.

The purpose must be to clean the horse's coat using brushes to remove the mud, dust, grease and sweat accumulated on the animal during the course of a twenty-four-hour period. It is vitally important that the exercise of grooming is carried out energetically because the horse will gain no benefit from merely being stroked. Apart from cleaning the coat and pores of the horse, grooming also serves to stimulate his circulation. The importance of keeping the pores clean cannot be over-emphasised because these tiny orifices allow the excretion of waste products from the body in the form of sweat. Sweating is a very necessary metabolic process acting to improve fitness and condition. For this reason clipping plays a vital role in removing the encumbrance of excessive hair when the horse is being conditioned. If the horse was unable to sweat freely, not only would his fitness be hindered but also his circulation would not operate with the efficiency necessary for the horse to thrive. As a consequence he could suffer undue effects internally because various organs are not working properly.

After using the body brush a damp cloth which has been rinsed in a bucket of hot water to which a handful of soda has been added, will pick up any remaining grease on the coat before a final polish with the stable rubber. It will not only save the tail from damage but also improve the look of the hair if a little bloom improver is applied before it is brushed out with the body brush. One of the best products is Absorbine which comes from the USA. A tail bandage used regularly while the horse is stabled will keep the tail lying flat and make it easier for pulling.

The grooming kit

Dandy brush. This is used for removing mud and sweat from the body and limbs of the horse. It should not be applied to the head because its long bristles are coarse and stiff; the horse will object to it and it could damage the sensitive skin. Neither should it be used to

Dandy brush.

71

brush out the tail as the bristles will damage the hairs and pull them out. The first stage of grooming is done with the dandy brush.

Body brush. The bristles of the body brush are softer and the shorter they are the stiffer they will be. After the surplus dirt has been removed by the dandy brush, the body brush together with the metal curry comb is employed to remove the grease from the coat.

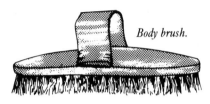
Body brush.

Curry comb. The metal curry comb with rows of teeth is used soley to remove the grease and hair which accumulates on the body brush as it is used. It should never come into contact with the horse because the serrated metal edge can damage the skin. The plastic and rubber types, on the other hand, are for use on the horse. Plastic curry combs often have sharp, pointed bristles and should be used with care as they will irritate horses with sensitive skin. The round, rubber type are less abrasive and often preferred.

Metal curry comb.
Rubber curry comb.

Water brush. Made of a soft bristle similar to the body brush the water brush is mainly used for damping down the mane and tail to encourage them to lie flat and before applying a tail bandage. It is very necessary in preparation for plaiting. A dry water brush can be used on the sensitive parts of the head where its uneven surface will reach some areas better than a body brush will.

Hay or straw wisp. As a cheaper alternative to a leather-bound banger, the home-made wisp serves to bang or strap the horse as a means of stimulating the muscles and thereby improving their tone. A hay wisp is used for banging while a straw wisp is often preferred for drying the horse. To make a wisp, take a large double handful of hay or straw and twist it at one end while holding the other end still by standing on it. After it is tightly twisted into a length of 18 ins–2 ft (45–60 cm) loop it around double and cross the lengths over so that it forms a pad.

Stable rubber. This is simply a cloth made of linen or cotton which is used to wipe over the horse after he has been brushed, to remove any dust. It serves to polish and effect a final bloom to the coat and should be used in conjunction with the wisp or banger to keep the hair lying flat between strokes of the wisp. Tea-towels are less expensive than stable rubbers purchased from the saddler and are just the same. Linen is preferred to cotton because it is stiffer and better wearing.

Mane combs.

Mane comb. A metal or plastic mane comb is used for combing out the mane in preparation for plaiting and for pulling both mane and tail. Traditionally metal combs are preferred to the modern plastic types although both types are suitable.

Hoof-pick. A strong metal hoof-pick is a vital part of the grooming kit. Its sole use is for picking out the horse's hoof. It is possible to buy hoof-picks with a small brush at the other

Hoof pick

end, which is preferable to having a separate brush to brush out the hoof. It is wise to tie a piece of string to the hoof-pick as this will make it easier to find if dropped in the bedding.

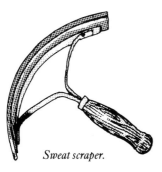

Sweat scraper.

Sweat-scraper. There are two types of sweat-scraper most commonly used today. One is the double-handed type which is a strip of flexible metal about 2 ft (60 cm) long, and the other is the half-moon scraper which has thick rubber on one edge and metal (or plastic) on the other. It has a short handle and can be used with one hand.

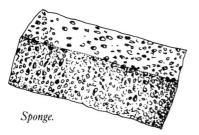

Sponge.

Sponge. A clean sponge is needed for wiping the eyes, nostril and lips of the horse; a separate sponge should be kept for sponging the dock and sheath. All these parts should be cleaned each time the horse is groomed, except for the sheath which needs doing once a week. A larger sponge can be kept aside for the job of sponging down after the horse has sweated. For the sake of hygiene a sponge which is used on the horse should not be used for anything else.

Banger. A leather or nylon pad can be bought as a ready-made banger in preference to making a wisp. Although they are obviously more expensive they last for many years providing they are looked after properly and kept supple with a leather preparation such as saddle soap or neatsfoot oil. They are only used for banging or strapping the horse and care should be taken to clean them between use on each horse to guard against the spread of disease.

Scissors. A pair of stable scissors should be included in the grooming kit for the purpose of trimming the bottom of the tail when it is banged, the top notch (between mane and forelock) and the inch or two at the bottom of the mane, i.e. over the withers. Custom-made stable scissors have a long blade which is curved upwards.

Cleaning and maintenance of equipment

At least once a week the brushes and curry comb should be thoroughly cleaned, not only for the sake of hygiene but also because there is no point in trying to clean a horse with dirty tools.

The brushes are best cleaned by dipping them in warm water to which is added a few soda crystals and using the hand to agitate the bristles. After rinsing in clean water the brush should be left to dry with the hairs down to prevent the base of the brush from rotting. The wood or leather base of the brush should not be wetted more than can be helped. The curry comb will require scrubbing with a hard brush in hot water and then left to dry with the teeth down to encourage drainage.

The mane comb should be brushed thoroughly between the teeth to remove grease and left to dry.

Stable rubbers will need washing at least every three or four days. It will be easier to put them in the washing machine than to wash them by hand.

Buying of grooming equipment

Economics play a large part in the selection of grooming kit although it must be said that anything but the best is false economy. Cheap foreign brushes will not last as long as they are made of inferior bristles which compose of vegetable fibres. Except dandy brushes all other superior quality brushes are made of animal fibres such as pigs' bristles and horse hair.

Quartering

This is a brief version of grooming, usually carried out at early morning stables after mucking out. The purpose is to make the horse look tidy before exercise. The method is to take the rugs from the front of the horse and brush off any stable stains and remove any bedding from the mane etc., then fold the rugs forward to cover the front of the horse and brush off the hindquarters, ensuring that no straw or shavings are left in the tail. A damp water brush can be applied to the mane and tail to flatten them down. The feet should be picked out, hoof oil applied and a tail bandage fitted. Rugs should be shaken thoroughly and individually before being put back on the horse. The horse can then be tacked up ready for work, or if he is not required straight away he can be rugged up again. Some owners prefer the horse to be left tied up once he is quartered to prevent him from lying down.

When to groom

The optimum time to groom a horse is when he returns to the stable after exercise as soon as he is dry. At this time the horse's pores will be open and therefore easier to clean. No attempt should be made to groom him until he is completely dry. It will benefit the horse to be groomed straight after exercise rather than after he has been standing. If it is not possible to groom immediately after exercise it should be carried out as soon as possible thereafter during that day.

How to groom correctly

First of all the horse should be tied up. Any rugs should be removed and hung over the door or laid in the manger (never on the floor). In the winter it will be necessary to cover half the horse with a rug whilst working on the other half, i.e. fold a rug across the quarters while attending to the front half and across the shoulders during the time spent on his quarters. The feet should be picked out into a skep and brushed with a hoof brush. If the horse has just returned from exercise his feet may have been washed, leaving only the hoof oil to be applied after grooming. The dandy brush should then be used all over the horse's body, beginning behind the head and working with the lie of the coat to remove any surplus dirt. Particular care should be taken where the horse has sweated and where mud is caked, especially on the limbs. Some thin-skinned horses such as Thoroughbreds will not tolerate the dandy brush and you may have to resort to the body brush to remove the more stubborn dirt. Horses who are ticklish and will not allow the dandy brush on their body may let you use straw to rub off any mud. As already stated the dandy brush should not be used on the face as the bristles are too harsh nor on the tail as it will damage the hairs and pull them out.

The second stage of grooming is the use of the body brush in conjunction with the metal curry comb. With one in either hand and beginning with the horse's head, the body should be brushed thoroughly and vigorously, again with the lie of the coat. Every half a dozen or so strokes the body brush should be cleaned with the curry comb. From time to time the curry comb will need tapping on the ground to remove the dust. To clean the body brush with the curry comb at almost every stroke is an unnecessary practice of idle grooms who are not necessarily engaged in good grooming. The body brush can also be used to brush out the tail but great care must be taken not to pull out hairs. The use of the body brush as part of grooming can only be effective if it is applied energetically by the groom. Mild strokes will only collect surface dust and not thoroughly clean the coat from the hair roots.

Time should be spent on the mane, brushing the underneath as much as the top side in an effort to remove scurf and grease from the roots. Too often today lazy grooms are less thorough than they might be when grooming manes and tails, and resort instead to the easier method of shampooing once a week, which produces an instant bloom to the hair and looks impressive. This practice may

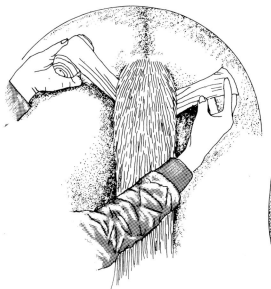

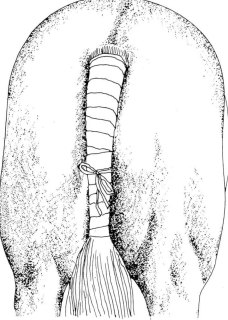

How to begin to bandage a tail.
RIGHT: *A correctly fitted tail bandage.*

be necessary for show animals but is generally overdone for other horses.

Once the body brush has done its job, if the horse is to be banged or strapped this is the time to do it. Horses who have not experienced banging must be introduced to it gently so as not to frighten them. A leather banger tends to make more noise than a hay wisp so the latter should be chosen for first-timers and youngsters until they become accustomed to it.

The four areas which are banged in an effort to stimulate the muscle and encourage development are the top of the neck, the shoulder, just behind the loins and the quarters. With a banger in his right hand and a folded stable rubber in the left, the groom should stand at angles to the horse with legs slightly apart in order to be balanced and able to apply prolonged effort. Beginning with the horse's neck the aim is to hit the horse with the banger hard enough for the horse's muscles to contract and in a rhythm which allows time for the stable rubber to be passed over that part in between strokes of the banger. It is important that the rhythm and pressure is maintained evenly to enable the horse to anticipate the

blow and thereby benefit fully. Be sure that the horse is standing square on each leg at all times otherwise the effect can be uneven. From the neck the groom should pass to the shoulder, then to the loins, and lastly to the quarters. The length of time spent on each part will depend entirely on each horse and the muscle development required in a specific area. The groom will have to use his discretion if the exercise of banging is to be fully effective without wasting time on well-formed muscles. On average, ten minutes on each part, i.e. the neck etc., will be adequate for most horses, up to a maximum total of one hour per horse.

Once the horse has been banged he can then have his eyes, nose and dock sponged followed by a complete wipe over with a damp cloth which has been rinsed in warm water to which soda crystals have been added. This will remove surface grease and dust before the horse is finally wiped down with a stable rubber. He can then have a tail bandage fitted (see illustration) and hoof oil applied, taking care not to get any oil on the horse's coat. Some people oil the inside of the hoof or part of it, but this is a matter of personal preference.

Hoof oil.

The rugs should be shaken and if necessary brushed off before replacing on the horse. Part of the grooming routine may include cleaning the roller. A leather roller will need washing before saddle-soaping and any brass buckles or rings should be polished. Webbing should be brushed daily otherwise manure stains will rot the material.

Mane pulling

From time to time it will be necessary to shorten and thin the horse's mane in order to keep it tidy, make it more manageable and enable the groom to plait up quickly and neatly. Some horses are left with their manes

in a natural state, such as Arabs, also some native breeds, but on the whole most horses have their mane pulled to about 5–8 ins (12–20 cm) long. The horse should be tied up and the mane thoroughly brushed out before being combed out with a mane comb.

The illustrations show how a mane should be pulled in stages and it is important to remember to take only a little hair at a time otherwise the horse will become sore at the hair roots and resent it. If the horse is having his mane pulled for the first time it is recommended that it is done in a number of short sessions taking, say, a little in the morning and a little in the afternoon for a few days until he gets used to it.

Laying a mane over

If a mane does not naturally lie flat to the off side it is good practice to either plait or bunch the mane over from time to time to encourage it to do so. Some horses may have just a part of the mane which lies on the wrong side and which may respond to a damp water brush wetting the upper part of the mane to flatten it. In the illustrations it can be seen how quick

A correctly pulled mane.

and simple a job it is to plait or bunch the mane. This is something that can be done after grooming, leaving the plaits in overnight.

To plait a mane over for this purpose requires only that the mane be plaited about half its length, as opposed to the whole way for dress. It will not be necessary to plait as tightly as one would otherwise do, and indeed it can damage the mane by pulling out hairs if tight plaits are left in for hours. The plaits can be much bigger too, and one does not have to be so careful about their uniformity. An alternative to plaiting over which can be equally as effective if the mane is short and tidy is to take bunches of mane over in small quantities, say, the width of a mane comb each.

Whichever way you choose first damp the mane with a wet water brush and brush it until it lies flat, then, starting at the top near the poll, take a section of mane by measuring it with a mane comb. Remember the individual amount of mane is not important as long as it lies flat. Plait or bunch the mane and secure it with an elastic band about an inch from the bottom. Continue in this way until all the mane is laid over the right side. When you reach the last one or two plaits over the withers be sure that the plait or bunch is quite flat because the rugs will rub them and possibly make the mane worse than it was before.

Great care should be taken when removing the elastic bands, not to damage the hair. Plaits or bunches will need to be taken out or redone after twenty-four hours because they will start to come undone on their own.

Plaiting a mane

For most competitive events and hunting it is customary to plait the mane and secure it with strong thread. The illustrations demonstrate the correct way in which this should be done. It is important to remember to take out the plaits as soon as possible after the competition or day's hunting to prevent damaging the hairs.

A time-saving exercise in preparation for plaiting is to cut a number of lengths of thread – which should be the same colour as the mane – sufficient for sewing one or two plaits at one time, and thread them onto needles. This will enable you to continue plaiting and sewing without stopping to thread needles. Be sure that the needles and thread are placed in a safe place, accessible to the groom but not in danger of falling into the bed, and out of the horse's reach. Some people prefer to use rubber bands for plaiting up, which they consider a quicker method. These will damage the hair if used regularly, give a less tidy appearance and are therefore not recommended.

To prepare for plaiting up you will need: a mane comb, six to eight short thick needles, a short pair of scissors, thread, a water brush and a stool on which to stand. The horse will have to be tied up and the front of his rugs folded back to expose all the mane. It will help enormously if the horse is not distracted by other activity in and around his box whilst he is being plaited. To effect a good job of plaiting the horse must keep as still as possible. Have all the kit you need in your pockets with the pre-threaded needles stuck in your coat on the front of your left shoulder or somewhere equally safe where you can reach them easily. Do not be tempted to stick them in the horse's rugs!

Some people prefer to plait the forelock first but traditionally you should begin at the poll and work down the mane, before finishing up with the forelock as the last plait. The number of plaits will depend on the thickness of mane but should add up to an uneven number on the neck. Seven or nine is an average number for most hunters although the modern fashion now seems to be to have many very small plaits rather than seven or nine larger ones. Much seems to depend on personal preference today rather than tradition, but if the latter principle is applied the type of event and horse will be considered. The size and depth of the horse's neck should also be looked at – a big heavyweight or cob can look very silly with tiny plaits. Show jumpers, dressage horses and many show horses are given several plaits which look smart if the job is well done, but again the number is purely cosmetic.

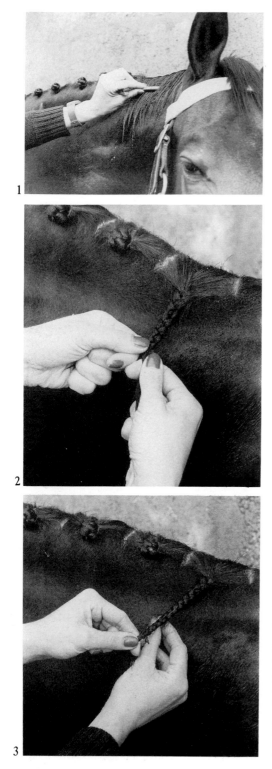

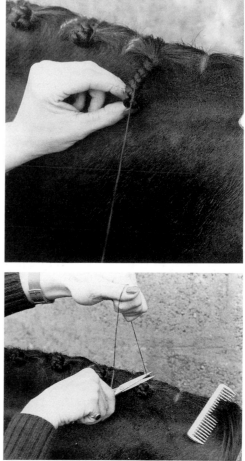

Plaiting a mane

Having brushed the mane thoroughly and combed it through, damp it down with the water brush, taking care not to wet it too much or it will slip through your fingers as you are plaiting. It is important that you maintain a firm grip on the mane to get a good, tight plait. With the mane comb divide a section of mane, not wider than the width of the comb, and push the comb into the adjacent remaining mane to keep it to one side whilst you divide the pieces of main into three even sections and begin to plait. It is critical to the final appearance of the plait that the start is tight and even. With each cross-over hold your thumb down on the top of the plait to keep it from slipping.

Continue plaiting until all the hair is plaited and hold the end firmly in one hand whilst taking the needle and thread to begin sewing up.

The length of thread should be double with the ends knotted together. Begin by bringing the needle up through the bottom of the plait (from underneath in the middle) and then wind the thread twice around the bottom of the plait, finishing up by pushing the needle down through the end of the plait and by folding the plait under itself to bring the end up to the roots of the mane. The needle can then be passed from the underneath of the base of the mane through to the top (upperpart) effecting a loop in the plait. Once this is pulled tight the plait can then be rolled under until it forms a tight ball at the top of the neck, and firmly tucked under the base of the mane. The needle can then be brought around and under the plait, pushing up through the middle and coming out at the top towards the root of the mane. Repeat this action to the left and to the right, each time pulling tight to secure the plait before cutting off the thread as close to the mane as possible.

Work down the mane in this way until all the hair is plaited. The forelock can then be plaited in the same way taking care to divide the mane into even parts and keeping it tight and square to the horse. The forelock plait should look central to the forehead and close against the head. If the horse is fidgety it will help to have someone hold him steady, particularly for the forelock plait. It is also advisable to undo a headshy horse from his tie-up ring before plaiting the forelock.

A worthwhile point to remember is never to wash a mane within two or three days of plaiting up because it will slip through your fingers and not hold into a tight plait.

How to undo plaits

It is important that plaits are undone carefully. A clumsy groom can easily spoil a mane which has to be plaited up many times in a season, by negligence with the scissors. Take a short pair of scissors and, always cutting with the line of

the mane, snip the thread at the top of the plait. This may need two or three cuts. The plait can then be unrolled and the thread at the bottom can be cut. The plait itself should then be undone with the fingers. Once all the plaits have been unpicked the mane should be brushed or combed out and damped down with a wet water brush to encourage it to lie flat. If the mane is not brushed out it will stand out from the neck and look as if it has just been permed. Do not leave the mane like this.

Hogging a mane

A hogged mane is one that has been shaved off at the root of the hair, removing the length of the mane and the forelock. Some establishments still prefer to see all manes hogged but, generally speaking, this is only done to polo ponies, cobs and some hunters. Clippers are the only efficient means of removing the mane tidily, and close to the roots. Care should be taken not to clip the horse's coat or cut the base of the mane accidentally.

It is unnatural for the horse to be without his mane and forelock which nature intended to serve as protection against flies.

Tail pulling

For many purposes it is preferable to shorten the top of the horse's tail by pulling it so that it lies flat and neatly against his dock. Racehorses are the exception to this as traditionally their tails are left in their natural state. No matter how well a tail is pulled the finished look will improve with the regular application of a tail bandage. It is recommended that the tail bandage be left on from the time the horse is groomed to late night stables, providing it holds in place. If it slips down it should be refitted throughout the day. For horses with untidy tails keep a tail bandage on throughout the day whilst the horse is stabled.

To pull a tail you will need only a mane comb, but first the horse must be tied up. The tail should be brushed and combed out at the top, i.e. the length of the dock. With one hand raising the tail the other hand is free to comb out and define those hairs on the sides of the dock which need pulling. If the hair is long it may be pulled in the same way as the mane, i.e. by 'back-combing' a piece of hair before pulling it out by its roots. On the other hand if the hair is short it may be possible to pull out the stray hairs using thumb and index finger or by gripping the hair between thumb and mane comb. Only hairs on the side and the underneath of the dock should be removed, never those on the top side. It is asking too much of the horse to tolerate long sessions of tail pulling; instead the groom should pull a little each day over a few days and be sure to keep a tail bandage on whenever practicable. The tail should, of course, be damped down with a wet water brush before the tail bandage is fitted to encourage the hair to lie flat. It is usually only necessary to pull out hairs to about 6–8 ins (15–20 cm) down the length of the tail. Great care should be taken not to make the horse's dock bleed unduly – some horses are more

sensitive than others and may only stand a little hair being pulled at a time before they start to bleed. A useful tip for pulling the tail of a difficult and fidgety horse who is likely to kick, is to have someone reverse him up to a stable door and hold him there whilst his tail is pulled by someone standing outside the box.

Banging a tail

This expression is misleading for it refers to something quite different from banging. All tails, whether or not they are pulled, should be banged at the bottom in order to cut them off level. The recognised length for the tail is level with the point of the hock when the tail is being carried in its normal posisition – that is to say, when the horse is moving, for it will be quite different from when the horse is at rest.

The tail should therefore be raised to its 'mobile' position by placing the forearm across and underneath the dock. The other hand should then run down the length of the tail until it reaches just above the point of the hock, bunching the hair tightly together in the hand. The other hand can then be freed from the top of the tail to take up a pair of scissors and cut off the excess hair. It is important to go through the tail with the fingers rather than a brush to separate the hairs before banging, otherwise you will have an uneven length.

Tail bandaging

In order to keep a pulled tail neat and tidy a tail bandage should be kept on as much as possible whilst the horse is stabled during the day. It should never be left on overnight because it could interfere with the horse's circulation. If the tension is not correct it will soon slip and do no good at all. It will eventually fall off and get into the horse's bedding when there is always a risk of the horse chewing it.

The illustration on page 75 shows how the bandage should be applied, beginning at the top of the tail, having first damped it with a wet water brush to encourage it to lie flat. With one hand holding the top of the tail to secure the

A correctly pulled tail.

bandage, the other hand should work around the tail to the bottom of the dock keeping an even pressure which is not too tight but will not work loose and cause the bandage to slip. The tapes of the bandage, which have been ironed flat, should then be tied flat into a conventional shoe-lace knot.

Plaiting a tail

For many different types of competition as well as hunting it is sometimes preferred to plait up a tail which has been left long in its natural state. If the tail has been pulled it will take some time before it is long enough to plait.

You will need a length of thread the same colour as the horse's tail. The tail should be brushed out so that the hairs are separated and the whole of the tail is clean without any bits of bedding in it. The horse must be tied up. Damp the tail slightly and take a thin group of hairs from each side underneath the top of the tail. Separate them from the rest of the tail and cross them over to start the plait. The tail is plaited in the opposite way to the mane. Continue to take small sections of hair from each side, plaiting them into the middle of the tail, and be sure that the plait is central to the tail all the way down. Keep the plait very tight from the beginning or the whole thing will look very untidy. Once you have reached the bottom of the dock plait up what remaining hair you have in your hand right to the end. Tuck the end of this long single plait underneath itself, taking the end back to the start to form a loop, then sew the end securely in place.

Tail plaiting needs a lot of practice because it must be tight from the top to hold firm and straight all the way down. A badly plaited tail is worse than no plait at all.

Singeing

Since the introduction of electric clippers singeing is becoming a less common practice. It is very difficult to acquire singeing lamps today so the usual substitute is a candle. The principle of singeing is to burn off the cat hairs of the horse's beard and throat. Its advantage is that it does give a more natural finish close to the horse and does not leave a clean-cut look which happens with clippers. Great care should be taken not to burn the horse's coat. Some horses may be startled by it so it is best to do it gradually to begin with. It is safer to do it outside the box in case matches or wax drop onto the bedding.

Trimming

Tails, fetlocks, ears, beard and muzzle are usually trimmed with scissors or clippers for the sake of appearance. Some people argue that the horse's whiskers around his muzzle and eyelashes should not be cut off because he needs them as feeling sensors. This is, of course, true but for cosmetic reasons they are often trimmed. The top notch, which is cut for the headpiece to fit between the mane and forelock, can be trimmed with scissors or clippers. It should be about 2 ins (50 mm) wide. The inside of the horse's ears can also be trimmed depending on personal preference. Show people tend to trim all the inside hair where others may just level the outside. Tails should be banged before being trimmed with the clippers or scissors. The beard can be singed or trimmed with scissors or clippers up the line of the jaw bone, taking care not to cut the coat. Some traditional stud grooms would always take off about 2 ins (50 mm) of the bottom of the mane but often the constant friction of the rugs will rub away some of the mane there. Fetlocks or feathers can be trimmed using a mane comb, held in the opposite direction of the coat to lift the hair to be cut with scissors. Alternatively clippers will do a tidy job providing the groom is careful not to cut into the coat and nick the skin.

Washing manes and tails

From time to time it becomes necessary to wash the mane and tail in order to remove the grease from the roots. It is best to choose a fine day without a cold wind to guard against the

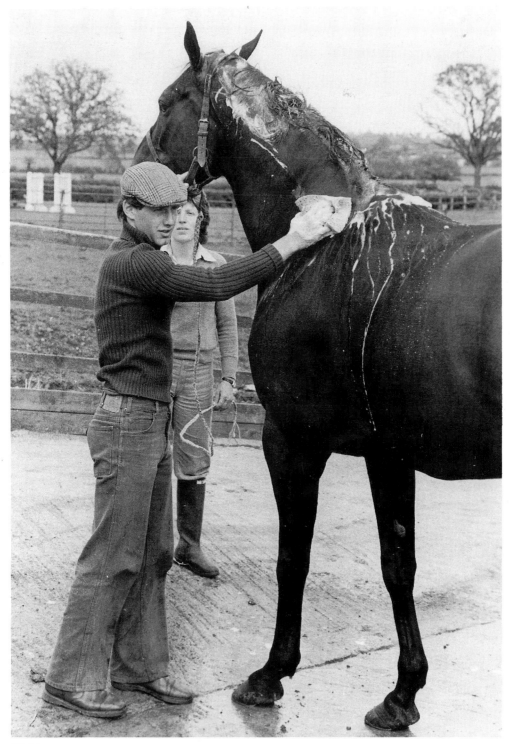

Washing a mane.

horse catching a chill. The most important thing to remember is never to wash or bath a horse who is even slightly unhealthy or he may catch a chill which can predispose more serious ills.

If you have a washing-down box so much the better if the weather is not very warm. Once the horse is clipped and in the winter months be sure to keep some clothing on his back during and after he is washed. Tie him up and wet his mane or tail thoroughly with water which should be made warm in winter but can be straight out of a hose pipe if the weather is warm and preferably when the horse is warm from exercise. Great care must be taken NOT to get water into the horse's ears as this will be the first thing to upset him. Apply a little animal shampoo and wash thoroughly before rinsing with plenty of clean water to remove every trace of soap. Use the sweat-scraper to scrape off as much surplus water as possible and swing the tail around for the same reason. If the weather is very hot the horse can go back

to his box but, generally speaking, it is best to walk him around to dry off before he is returned to his stable. A tail bandage can be fitted when the tail is still wet taking care not to bandage too tightly because the bandage will get tighter as the tail dries.

Bathing and drying a horse

In Britain it is only usually warm enough to bath a horse completely in the summer months and even then one would recommend that it is done after the horse returns to the stable still warm from exercise. He will benefit mostly at this time because his pores will be open. A bath will also refresh him just as a shower after a game of tennis would do us. Choose a sheltered area to tie the horse up, outside if you don't have a washing box. Wet him completely and then apply shampoo. Rub him all over thoroughly, making a lather, paying particular attention to the back. Once the horse is

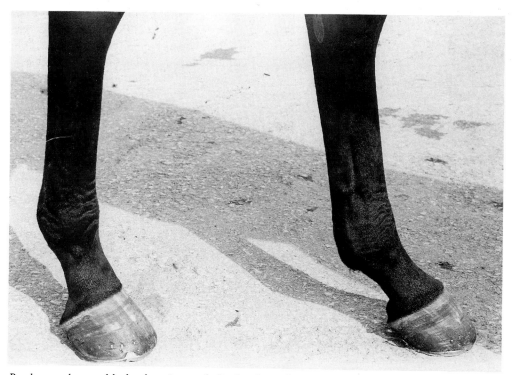

Bandage marks caused by bandages incorrectly fitted or for too long or whilst the coat is still wet.

washed he can be rinsed off and scraped with the sweat-scraper. A sweat sheet should then be put on and the horse walked until he is dry. Depending on the weather it may be necessary to put on a towelling sheet or similar light clothing to guard against the horse catching a chill. Some people put the horse on the lunge and jog him gently until he is dry but on no account should he be put back in his box until he is quite dry.

Only if the weather is hot should cold water be used, otherwise it should be warm. If the horse shows any sign of not being warm when he returns to his box he must be rubbed down energetically until he is quite warm and comfortable. If his legs are still not dry bandage him with woollen bandages and be sure he has enough clothing. Close the top door if there is any risk of a draught.

Mechanical grooming

It is becoming increasingly fashionable now, particularly in large stables, to use electric groomers as a time saver. There are three main types on the market, one of which is the vacuum system. This has a cylinder similar to a household vacuum to which is attached a length of plastic hose. A rubber curry comb is then fitted which has a hole in the centre to suck in dirt and dust. It is particularly useful for removing surface mud and dust and also in the springtime when the horse is losing his winter coat. It can be coordinated with conventional grooming most conveniently for horses who are not in hard work or who are unclipped.

The other types are a rubber vacuum which is fitted with a rotary brush and a rotary type which has removable heads for body brushing or dandy brushing. These types are very good, especially the body brush, on a clipped horse. As well as keeping the coat clean an electric groomer massages the skin better than one could by hand. Generally speaking if the horse is introduced to the groomer carefully he will not object but instead find the experience invigorating. The groom must be particularly careful when grooming near the mane and make a habit of bandaging and tying up the tail. Never use the groomer on the horse's head. Once the horse has been thoroughly groomed he should be rubbed down with a stable rubber. The horse should always be tied up for electric grooming as a safety measure because of the electric cable. Ideally a separate box could be set aside in which to use the electric groomer and for clipping and washing.

Washing the genitals

A gelding will need his sheath washed in an extended position as a matter of hygiene. This can be done once a week with warm water using carbolic soap. If the horse objects have an assistant hold him with a front leg held up. Use plenty of water to rinse him clean. It is worthwhile getting the horse used to this at an early age.

The mare's dock and vagina will also need the same attention but more frequently, especially when she is in season. Care should also be taken to wipe the under part of the tail on all horses because this can also become dirty and if neglected may become sore and the skin will peel. If the tail bandage is too tight it will cause the tail to become sore underneath and hair may fall out.

CHAPTER TEN

CLIPPING

Reasons for clipping

Primarily it becomes necessary to clip a horse when an excess of coat is proving a hindrance to him. Only horses in work need their coat removing in this way. A horse at rest needs his coat for protection whereas a working horse is likely to be stabled and therefore benefitting from artificial protection from the weather. A horse who is worked with a winter coat will sweat excessively and be impossible to clean and dry properly even after a small amount of work. His pores will be clogged and unable to breathe, and he will be susceptible to colds and chills. Condition will soon be lost if he is worked with a winter coat.

The horse should be brought up and starting work two to three weeks before his first clip. The more commonly bred the horse the more coat he will grow and it may therefore be necessary to clip him right out at the first clip. Afterwards his legs and a saddle patch can be left on to give him some protection. With Thoroughbreds it is usual to leave the saddle patch and legs unclipped throughout because their fine coat will not hinder them.

The summer coat will begin to fall out during the autumn and the first clip is normally done each year about October time (traditionally referred to as blackberry time), once the new coat is completely through and 'set'. By then it will be easier to clip and not require doing as often as a coat which is clipped before it has finished growing. After the initial clip and depending on the type of horse and the work which is required of him, he may be clipped every three to six weeks. After Christmas it will be found that the coat does not grow as quickly. Instead, what are known as 'cat hairs' will grow; the more common the horse the more apparent they become. They are coarse, irregular hairs which soon give the horse an untidy appearance if they are not trimmed regularly. Care should be taken as spring approaches not to clip the summer coat when trimming cat hairs. Alternatively a singeing lamp or candle can be used to remove cat hairs on the legs and under the jaw and neck, as described in Chapter 9.

Horses which come into work in the spring or late winter-time can be clipped out providing the coat is not changing rapidly. During the spring and summer months many competition horses are clipped to allow them to sweat more freely and keep them cleaner in their coats. Show horses such as cobs and show ponies are often kept clipped right out to enhance their profile. Greys are particularly improved by continual clipping but chestnuts tend to look very pale. Clipping the horse's facial whiskers is mostly a matter of preference. Although it is necessary for showing it must be remembered that they are a sensitive organ of touch for the horse. Ears, too, are sometimes left unclipped by choice but the edges can be clipped to give them a tidy finish. The hair inside the ear gives the horse protection from flies in the summer as well as from the cold and wet. It is possible to buy coarse blades which only clip half the thickness of coat. These are useful for legs and for trimming a horse with a coarse summer coat. Clipping legs, however, must be a matter of opinion and preference: on one hand the coat protects the horse from thorns and mud but is consequently more difficult to clean and dry, while on the other, the clipped leg makes dealing with cuts and thorns a lot easier.

Whatever job the horse is required to do, once he is clipped and in work grooming is much simpler and the horse can be kept cleaner and dried more quickly. Special care must be taken to thoroughly clean and dry the legs, especially the heels of clipped legs as they are more prone to mud fever and cracked heels, both of which can render a horse quite lame and sometimes keep him out of action for several days. Always bandage clipped legs after hunting once the legs have been dried. The unclipped saddle patch will offer some protection to the horse from sores and scalding but the common horse is better clipped out especially if a numnah is used, for a wet coat will take a long time to dry. It may be necessary to clip a horse which has lice in the spring, or ringworm; he can then be shampooed and more efficiently treated.

Types of clip

There are a few different types of clip from which to choose depending on how much coat you wish to remove. This will be based largely on the type and frequency of the horse's work, his environment, the type of horse (i.e. Thoroughbred or less well bred), and to a certain extent your personal preference. Whichever you choose it must primarily be suitable and practical for that individual horse and you must be prepared to afford the necessary compensation, i.e. stabling, clothing etc., which he will subsequently need.

Full clip

This clip should only be used on horses who are in continual hard work and stabled for twenty-four hours with sufficient clothing. It involves clipping out the horse completely. Police horses, army horses, show jumpers, hunters, show horses and ponies are generally given a full clip. There is a risk of cracked heels and mud fever in the winter months if great care is not taken to dry the legs thoroughly, especially the heels, each time the horse returns to the stable after work. It is recommended that new blades are used for a full clip. Take care not to cut the skin with new blades where it is uneven and difficult to achieve a flat surface.

Trace clip

This is a particularly useful clip for ponies in the winter months and for young horses who are not in fast work. It means that the back, loins and legs are still covered while the parts most likely to sweat are clipped out and therefore easier to dry and groom. There are two variations on this clip depending on the work the horse is required to do, his type and, as

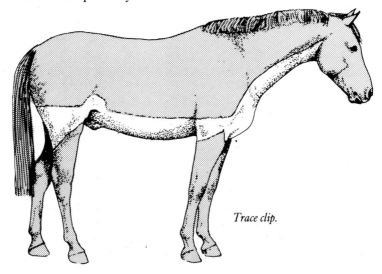

Trace clip.

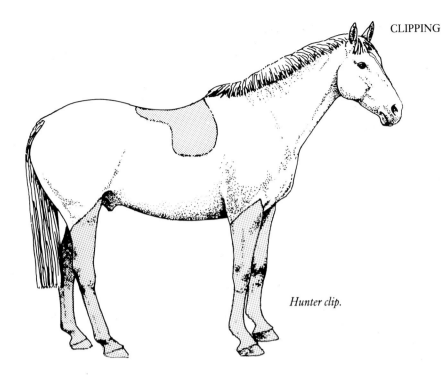

Hunter clip.

always, his environment. The choice of a trace high clip and trace low clip will also depend a little on personal preference. This clip is especially convenient in that the horse can still be turned out with a New Zealand. The lines on each side should be kept level and straight otherwise the whole thing will look very untidy. A variation of this clip that is useful for young horses in light work is when the whole of the quarters are left unclipped and a straight line is taken from the horse's stifle straight to his ears and from there down to the corner of the mouth.

Hunter clip

The hunter clip involves removing the entire coat of the body, possibly with the exception of the saddle patch, although this is sometimes removed on the first clip of the season and subsequently left to offer the horse some protection against saddle sores. The same applies to the legs which, again, are a matter of preference. While some argue that the horse needs the protection of a longer coat against thorns, cuts, cracked heels and mud fever others prefer the horse without, which certainly makes for easier grooming. Clearly, if

the coat is removed the horse must be kept warm with woollen bandages especially after he has cooled down from exercise. As with the saddle patch the legs can be clipped at the first clip and thereafter left on. Leg blades clip less close to the skin, leaving some coat but still looking tidy. The angle at which the tops of the legs is clipped is fairly critical because it can spoil the whole effect if it is not done to suit the horse's conformation. Never clip too low down as this will make the horse look very leggy. This is particularly relevant to the hind leg. The illustrations show how important this is.

When preparing to clip around the saddle patch, fit a saddle without a girth or stirrups and use a bar of saddle soap or a piece of tailors' chalk to mark a line around it with an inch to spare so that the saddle can then be removed and the area clipped equally on both sides. To achieve a level line at the top of the forelegs on both sides of the body take a piece of string and measure from the withers. Check the line is level by standing in front of the horse and stooping down so that you can assess it in relation to the rest of the body. Similarly with the hind legs, stand directly behind the horse and hold the tail to one side while you judge the level and angle with your eye.

Blanket clip
A blanket clip is just as the name suggests. Only the coat in front of the withers and under the belly is clipped leaving the horse's back and legs unclipped. Again, preference will play a part as to how much is taken off below the level of the saddle flap. This is a clip which is favoured for racing horses, point-to-

pointers and some eventers as it leaves the horse's back covered which is especially useful for cold-blooded horses.

A useful clip for horses who are only doing very light work but are beginning to sweat and will therefore benefit from having just a little coat removed, is to take away the hair from under the neck, between the front legs and under the belly.

Blanket clip.

Racing Clip. This clip is also useful on other competition horses particularly thoroughbred types who need to keep some coat on their back.

Preparation and introducing a horse to clippers

Before beginning to clip, the horse should be groomed to ensure that his coat is clean and free from any sweat, mud, grease or dirt which could interfere with the proper working of the clippers. The clippers cannot operate efficiently and the mechanism will become damaged if they are allowed to become clogged with grime. Equally important is the need to ensure that the clippers do not pull at the horse's coat and make it uncomfortable for him as this will cause him to be intolerant. If the weather is suitable it would be wise to shampoo the horse the day before he is to be clipped (this only applies to horses who have been stabled for a time beforehand). Never attempt to clip a horse who has come straight out of the field. He should preferably have begun walking exercise so that his pores are active and he is being groomed.

The young horse which has never experienced clippers must be gradually introduced to them whenever possible. It is worthwhile switching on the clippers outside a youngster's box so that he becomes accustomed to them from an early age and, day by day, one can take the clippers nearer and nearer until he is confident enough to allow the clippers against his shoulder and feel their vibration. The length of time it takes for the horse to be relaxed enough for you to get this far will depend entirely on the animal and the amount of handling he has already had. As with any form of education for the young horse, time and patience are essential and a bad experience must be avoided. In some cases it may be possible for a young horse to hear another being clipped, and such opportunities can be very useful. Whatever stage of learning the horse is at be sure never to switch on the clippers whilst you are standing near to him but stand away and give him a moment to get used to the noise. Make a practice of speaking to the horse all the time he is being clipped. During clipping take a break from time to time and give the horse a rest. This will also give the clippers a chance to cool down.

Before you begin to clip have someone hold the horse with a headcollar or bridle. Only mature, experienced horses who are quiet to clip may be safe enough to tie up and clip singlehanded. Otherwise, have an assistant hold the horse whilst he is being clipped. Do not leave anything to chance or expect a horse to stand like a lamb even if he is otherwise quiet to handle.

Accidents happen all too often with animals and there is no room for complacency at any time. Even an experienced person cannot clip a horse efficiently if the horse is constantly fidgeting. Some horses settle better if they have a haynet to occupy them, particularly when you are clipping the more sensitive parts. Special care must be taken during clipping not to catch the mane, forelock or tail with the clippers. If this happens and a piece of hair is removed from the mane it will look particularly unsightly and will stick up while it is growing back. You should aim to carry out the job smoothly and without causing undue stress to the horse, effecting a good clip in the minimum of time. The more you delay the more bored the horse will become, so try and do the job as quickly as is practical without unsettling the horse. However, the job should never be hurried and you may need to allow extra time for temperamental and young horses.

Safety must be a priority at all times. You are advised to wear rubber-soled footwear when clipping and ensure that you do not have any loose-fitting clothes or long hair hanging which might become tangled with the clippers. An overall or boiler suit, plus a hat or headscarf, will help protect you from what is a very dirty job. The cable of the clippers should be kept away from the horse and out of your way too. Beware that the horse does not tread on the cable and cut through it, which is more likely if he is wearing shoes. Make sure that you stand between the horse and the cable and socket. If the horse is clipped in his box, the bed should be put up and the floor swept clean. Ideally a separate box which has no bedding at all can be used. Good light is essential if there is any hope of doing a good job otherwise it may be found that patches have been missed once the horse is brought into daylight. Before you start close the stable door behind you and check that the cable is not in anyone's way. Before reaching this stage you will have already decided on the type of clip you are going to employ. In the case of the young horse who has not been clipped before it may be prudent to only do a trace or blanket clip which does not involve too much clipping for the first time. It is advisable to start the clip at the shoulder area and work towards the head. In this way if the horse proves difficult you will not have committed yourself so finally as if you had begun on the quarters. It will also give you some idea as to the horse's reaction and behaviour. At all times it is important to keep the horse quiet and relaxed because once he becomes upset and starts to sweat it will be impossible to clip him. Even a sedated horse can sweat and make it very difficult for the job to be done.

Many horses who are not used to clippers are apprehensive about being clipped round the head. It will be necessary to exercise a great deal of patience and, at the same time, firmness. An assistant is particularly useful here because you may have to stand on a stool to reach. At all times, and especially whilst you are clipping around the head, beware that the nervous or bad-mannered horse does not strike out with a fore leg; likewise if he is prone to kicking or biting you will have to be constantly on the alert and prepared for this. Rather than trying to clip the head whilst the headcollar is still fitted you will do a better and quicker job if it is removed and fastened around the neck with the assistant standing on the opposite side to you and holding the head-collar, with one hand on the horse's nose to keep him still. Start by clipping the cheekbone and if the horse objects very strongly you can avoid clipping his face and simply take a line up the side of his head to the base of his ear.

Care of the horse during clipping

If possible, avoid clipping on a cold and wet day when the horse would be more likely to suffer from the loss of his coat. Ensure that during clipping the horse is protected from draughts. As the front of the horse is being clipped cover the quarters with a rug and/or blankets. Likewise, the neck, shoulders and as much of the back as possible should be kept covered whilst the quarters are being done.

The use of restraints

Some horses will not tolerate clipping of their more delicate parts, e.g. belly, head and inside hind legs, and it may be necessary to restrain them in some way. Whatever technique you use remember to try and keep the horse as relaxed as possible at all times. The different methods which can be applied are: (a) holding up a leg; (b) pinching a piece of skin on the neck; (c) a nose twitch; (d) tranquillisers via intra-muscular or intravenous injection; (e) hobbles. It is recommended that these methods are resorted to in this order, i.e. if the first is not effective try the second, and if the second does not work try the third. The fourth option can then be applied but not before a considerable effort has been made on the part of an experienced handler to try and do the job using less drastic means. It may be necessary to solicit the help of a third person when using a restraining method so that one person is free to concentrate on holding the horse. Much depends on the nature of the horse and the experience of the handlers.

(a) The first choice of holding up a front leg will often work on quieter horses. Care must be taken by the assistant not to get in the way of the operator and to keep a firm hold of the horse.

(b) The second method of pinching a piece of skin can often be extremely effective for subduing a restless horse. Take hold of a piece of loose skin between thumb and forefinger from the middle of the neck and hold it. If the horse does not settle twist it slightly and if this still does not work take hold of a bit more. Once you let go, rub the area to prevent it from bruising and swelling.

(c) More difficult horses will need the nose twitch. This is by far the kindest and least stressful means of subduing a horse. The assistant should be experienced in the use of a twitch and use both hands to hold it firmly in place once it has been applied to the horse's nose. He should be aware of the hazards of letting go if ever the horse should become awkward and throw himself about, because the twitch will unwind and frighten the animal as well as endangering both horse and people. Generally speaking, horses who object to clipping do so mainly whilst being clipped on the head or any ticklish parts, such as under the belly, the elbows and between the legs. Be sure that no one stands in front of the horse while he has a twitch on because he is likely to strike out. The twitch should only be left on for as long as it takes to do the job and then be swiftly taken off, i.e. unwound quickly. This is often when the horse will throw his head about to shake off the twitch as it comes loose. The use of a twitch on the ear is not a means of restraint which the authors recommend because they consider it unnecessary and cruel. It can also make the horse permanently headshy.

(d) An alternative method of restraining a horse is to sedate him. For this the veterinary surgeon must be called in and he will advise and admininster an appropriate tranquilliser. There are numerous types available but ACP Pethidine and Rumpen are widely used for this purpose. Horses respond differently to tranquillisers and some may not react favourably. It is therefore essential that the vet supervises their use in case a sedated animal throws himself about in a dangerous manner.

It may be necessary to immobilise the horse completely and clip him on the floor in which case the authors recommend that he is taken to the surgery and the job carried out by professionals. If the animal is sedated in his box at home ensure that he has considerably more bedding than he otherwise would to protect him while he is on the floor and when he is regaining consciousness. As he recovers from the drug it is advisable to have someone holding his head whilst another holds his tail straight out behind to help him regain his balance. There is a danger that in stumbling about in a 'drunken' stupor he could fall and break his back, so every effort must be made to keep him straight and encourage him to stand still for as long as possible. The sedated horse is prone to breaking out in a sweat, which makes clipping near impossible. It is recommended therefore that two people clip the horse at the same time to expedite the job. As the horse sweats and during his recovery from sedation, be sure that he has sufficient clothing to keep him warm because he will be particularly vulnerable to catching a chill. A sweat sheet, under which has been placed some straw over his loins, with a jute rug upside down on top may be enough, but as the horse dries have some blankets ready in case he needs more protection. Each horse will react differently to sedation and restraints and it is recommended that only experienced horsemen/women use them and that any assistants are well briefed as to the possible consequences.

(e) Finally, hobbles can be fitted to immobilise the horse. It is, however, recommended they are only used in conjunction with tranquillisers when a veterinary surgeon and experienced handlers are on hand to help. There are different types of hobbles available but all do the same job, being fitted to the legs usually around the pasterns. A rope is then passed through them and all the legs can then be brought together which renders the horse quite immobile so that the handlers (in this case those who are to do the clipping) can work unhindered. Once one side of the horse has been clipped the animal can be turned over by pulling on the rope from the other side of the body and lifting the legs over while someone else turns over the head.

Care of the clippers

Before you begin to clip you must ensure that the clippers are in good repair and safe working order. The blades will need re-sharpening from time to time and it is therefore important to see that this is done promptly and that blades are not left lying about where someone could pick them up believing they are fit for use. To try and clip a horse with blunt blades will risk hurting him as they will pull on his coat. They must be kept as clean as possible during use by brushing off the hairs with a dandy brush and oiling with either 3-in-1 oil or a similar thin oil. The air filter must be regularly brushed off and kept clean to avoid over-heating. An inexperienced person should be taught not only how to clip efficiently but also, equally important, how to maintain the clippers and ensure that they run smoothly. Many mechanical faults are due to the lack of maintenance so for safety and economics (clippers are not cheap to buy or repair) make sure you look after them.

Before clipping begins it is important to establish the correct tension on the blades. Manufacturers do offer guidance which should be followed because incorrect tension will not only affect the cutting of the coat, i.e. it will not cut cleanly and efficiently, but will also risk damaging the machine because it will over-heat if the blades are too tight. Likewise, loose blades will not cut properly but make a mess of the horse's coat; the hair will get between the blades and jam them. Always ensure that you have a spare set of blades before you begin clipping in case of wear or damage to the others. If, during clipping, you find that the blades need more tightening it may be that they need resharpening. A set of blades will normally clip two to four horses depending on the thickness and state of the coat. If the clippers become hot during clipping switch them off and allow them to cool down before re-commencing. It is advisable to keep the clippers and their blades safely in a box which can be carried about.

Any repairs and servicing of the clippers should be carried out by the manufacturers or their recommended agents. It is both unsafe and false economy to allow anyone else to interfere with them because manufacturers will not support a warranty whose terms have been broken by the consumer. Damage to component parts will prove expensive enough if they are not under guarantee. Half the skill of clipping lies in the smooth running and proper maintenance of the clippers.

Choosing clippers

Before you make your choice of clippers you will need to establish just how much clipping you anticipate doing in the course of a year and also who will be operating them. A lightweight pair will suit the lady owner with one or two horses but would be less suitable for clipping a number of horses regularly, say, for hunting throughout the winter. In this case a more powerful heavy-duty type of machine would be needed. In a large busy yard it is advisable to have two pairs working efficiently so that if one pair breaks down you will not be left with a half-clipped horse. Although the initial expense of buying two pairs may be high it will soon be justified in an emergency, and providing they are properly maintained and serviced they should last for many years.

A number of people now opt for mini-clippers (otherwise known as dog clippers) for trimming. Being battery-powered, they need re-charging regularly and should not be used if they start to fail, even slightly, as they will pull at the horse's coat. Although they are expensive they make hardly any noise at all which makes them ideal for young or nervous horses. One should not, however, try and clip a horse out with them because they are not designed for this purpose. Horses, particularly those with a nervous disposition who would not otherwise tolerate being clipped around the head, will often suffer battery-powered clippers quite happily because of the noise reduction. A less-expensive alternative is hand clippers which, again, are only designed for trimming. They resemble a barber's trimmer and need to be used fairly fast and evenly to avoid pulling the hair.

Gone are the days – thankfully – of the old-fashioned wheel machine which needed two people to operate it, one to turn the wheel, (which had to be done at a regular pace to achieve an even clip) and the other to do the clipping. As one can imagine, this was a laborious job which could take up to two or three hours. Another obsolete machine was a motor-driven cable which turned the head of the clippers. Similar to a sheep-shearing machine, suspended overhead, these were particularly hard-wearing and therefore useful for clipping large numbers of horses.

A word of warning on buying clippers: avoid secondhand machines as they will not be covered by a warranty. It will be difficult or impossible to establish their history, i.e. service record and age. If and when they do have to be returned to the manufacturers it will doubtless prove to be an expensive operation without a guarantee or proof of purchase from a recognised dealer.

How to clip

Be sure that the horse is well groomed, with no wet patches, mud or excessive grease before you begin to clip. Always start clipping at the shoulders which is a less sensitive area and less likely to alarm a young horse. Ensure that the blades are flat against the horse's coat and not pressing into the skin, and that the blades are not pointing into the skin which can wrinkle and cut. Using long strokes with even pressure let the clippers ride forwards against the coat. After the initial stroke overlap the coat by half to threequarters of the width of the blade. If you try to cut more than this at a stroke you risk clogging the blades, especially if the coat is thick. Also you are less likely to miss hairs or whiskers if you go over the area twice and there is less chance of lines too, if an even pressure is applied in this way. It is always much easier to clip a horse the first time when he has a full coat than in subsequent clippings when the coat is much shorter.

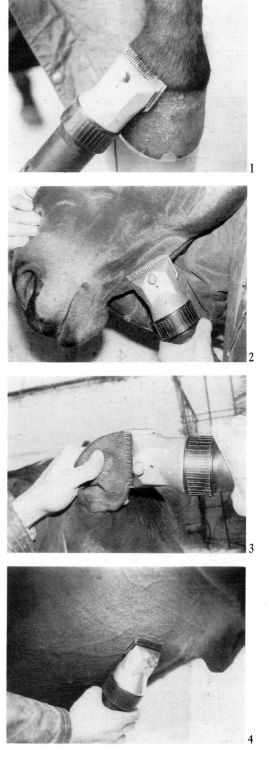

1 *The leg should be held up to be clipped below the fetlock.*

2 *Particular care must be taken on the head to clip against the lay of the coat.*

3 *Hold the ear so it is flat for clipping.*

4 *Bend down underneath the horse to ensure that the clippers are flat against the lay of the coat when clipping the belly as the coat goes in different directions.*

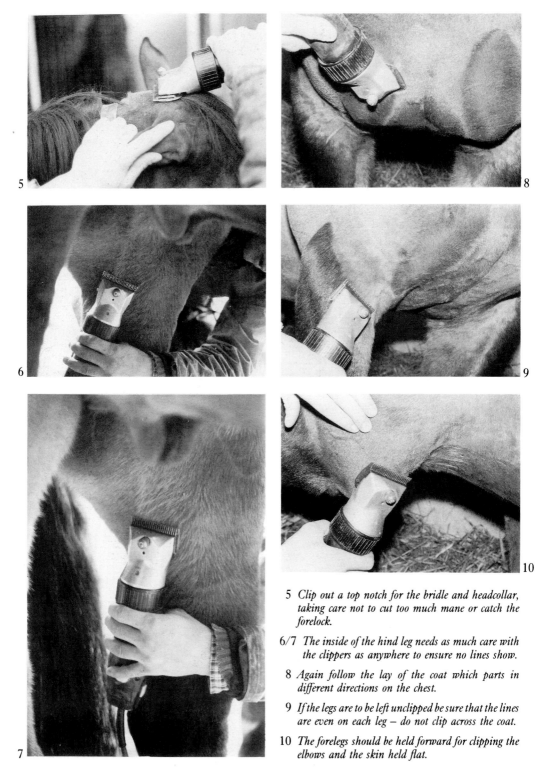

5   Clip out a top notch for the bridle and headcollar, taking care not to cut too much mane or catch the forelock.

6/7   The inside of the hind leg needs as much care with the clippers as anywhere to ensure no lines show.

8   Again follow the lay of the coat which parts in different directions on the chest.

9   If the legs are to be left unclipped be sure that the lines are even on each leg – do not clip across the coat.

10   The forelegs should be held forward for clipping the elbows and the skin held flat.

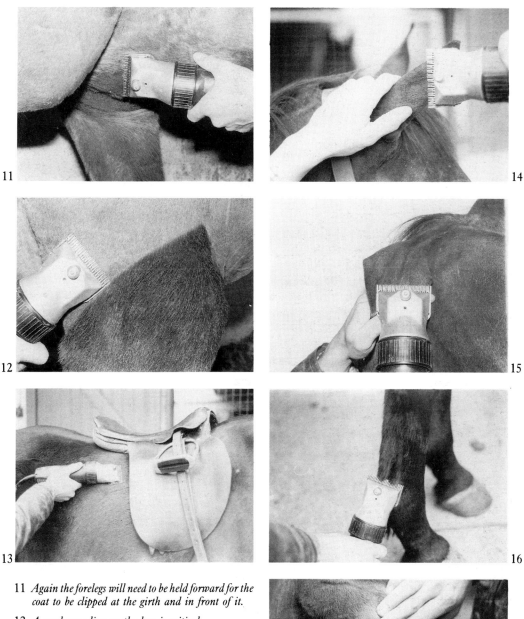

11 *Again the forelegs will need to be held forward for the coat to be clipped at the girth and in front of it.*

12 *A good even line on the legs is critical.*

13 *Fit a saddle to mark the saddle patch for clipping accurately.*

14/15 *Hold the ear flat in one hand.*

16 *Always clip up and not down the leg.*

17 *Stretch the skin flat with your free hand.*

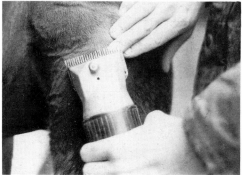

18

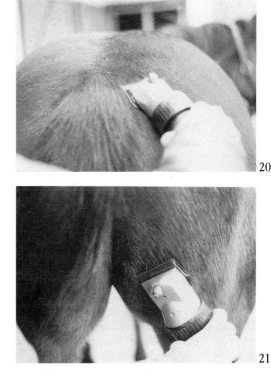

20

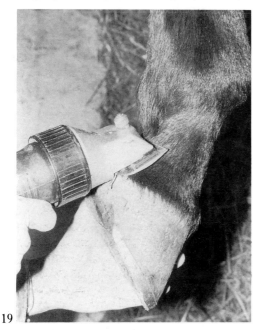

19

21

18 *Move the mane out of the way for clipping under-neath, taking great care not to clip too near and catch it in the clippers. (If the mane is caught it sticks up and looks very untidy for a long time).*

19 *Hold the leg up to clip below the fetlock.*

20 *The v at the top of the tail should be clipped evenly.*

21 *Hold the tail to one side and clip inside the hind legs.*

Continue to clip quickly and quietly with a minimum of fuss and distraction. If the blades overheat, switch off the clippers from time to time to allow them to cool down and oil them. Be sure to wipe the surface of the blades quite dry before using them on the horse again or the oil will smear the coat, making it difficult to clip, and could easily blister a horse with a sensitive skin. Some lines may appear immediately but these may be caused by pressure to the skin and, if so, will disappear within a day or so. Some horse's skin reacts to being clipped better than others. Sensitive skins tend to wrinkle and twitch while others are able to tolerate the action of the clippers

anywhere on the body. Care should, however, be taken on any horse when clipping the elbows and similar areas of loose skin which wrinkle easily. The skin should be pulled flat with one hand to stretch it as flat as possible.

It is a matter of preference as to whether the head is clipped first or last and much will depend on the horse. Once the neck and shoulders are clipped the body and legs can then be done. The chest is an awkward area to clip as the hair lies in different directions. The clip line on the top of the legs should follow the line of the muscle. It is important to take a positive direct line here and it must be level with the opposite leg; to deliberate often results in uneven lines. Care must be taken when clipping the base of the mane and it is better to be cautious and leave a margin for error. Nothing looks worse than a ragged edge where the blades have caught the mane itself; moreover, this can take months to grow out during which time it will stick up and be difficult to plait. In the case of a full or hunter clip the top of the tail should be clipped to an upside down 'V' shape which should neither be too small nor too large but must be straight and even.

When clipping the head have an assistant on the opposite side of the horse with one hand on the nose and the other on the poll, avoiding holding the ear if the horse will keep still. Your spare hand will then be free to cover the horse's eye while you clip the bony structure of the face, which needs careful attention. When clipping the ears hold the ear down so as to enable you to keep a flat surface, with your hand underneath for support. Again care must be taken to hold the forelock out of the way and not catch it with the blades. This can be done by the assistant. Whorls may have to be clipped from various directions according to the lie of the coat, to achieve a good result. Before you finish and rug the horse up it is prudent to brush over the horse to check for any hairs which may have been missed.

Some yards have a box specifically for clipping but if this is not possible ensure that you have a clean dry floor on which to stand the horse. Good lighting and an electric socket nearby are priorities. Never attempt to clip a horse on deep litter with poor lighting. Shadows will distract from clipping well.

Care of the horse after clipping

No matter how clean you thought the horse was before you started clipping it will be found that there is still much dirt and grease left in the coat after you have finished. Depending on the weather there are different ways of cleaning the horse after clipping. If the weather is warm enough and the horse can be lunged quietly to keep his circulation going, then the best way of giving the horse a thorough clean is

to shampoo him. This applies to horses with a full or hunter clip and would not be practical with any other type of clip, although clipped areas could be sponged off with warm water. The horse could in fact be ridden after a shampoo providing his saddle area is dry enough to put on a numnah and saddle, but whatever method you use, especially if the horse is just walked out in-hand, be careful that he does not catch a chill. He could wear a sweat sheet and/or towelling sheet depending on the weather, but he must be dried thoroughly before he is rugged up again in his stable. If the weather is not conducive to shampooing the horse will have to be groomed and strapped. This can be made easier if, after brushing, the clipped coat is wiped thoroughly with a cloth which has been rinsed in warm soda water.

Rugs for clipped horses

The type of rugging the horse will need will depend on the time of year, the weather, environment, type of stabling, the type of clip and not least the type of horse, i.e. cold- or warm-blooded. Whatever the situation it is good practice to fit a cotton or linen sheet first which will save the blanket and rugs from any grease and dust in the horse's coat and can be washed regularly. Also, a woollen blanket can cause a certain amount of irritation, especially on a horse with a sensitive skin. Generally speaking, the horse with a full or hunter clip will need two blankets and a night rug. If the legs have been clipped out it may be necessary in cold weather to fit woollen stable bandages for extra warmth.

If he is to be turned out he will need a New Zealand rug, worn with a blanket underneath in extremely cold weather.

During exercise a quarter/exercise sheet under the saddle may be needed to keep the horse's loins warm and to protect from the rain. In the case of hunters and polo ponies where a horse is led off another and does not have a saddle on, he will need a rug and roller/surcingle with a breastplate fitted to the roller/surcingle to prevent it from slipping back.

CHAPTER ELEVEN

SADDLERY AND CLOTHING

Bits and bitting

Bits are an important means of guiding the horse both on the ground and from the saddle. Bitting is virtually an art and certainly considered to be a controversial subject. Horses are different in size and shape; some have deeper or shallower mouths than others and the thickness of the tongue may also vary. A horse with a thin tongue often accepts a bit with a straight mouthpiece but, as a rule, straight mouthpieces cause discomfort by placing pressure on the tongue. Jointed bits will relieve pressure on the tongue and are suitable for the majority of horses, but there are always exceptions. Unsuitable bitting can ruin a horse's mouth permanently just as a rider's bad hands can undo years of careful, sensitive bitting. The less capable horseman should be given the mildest bit to safeguard against harming the horse's mouth. If a horse is over-bitted (that is, fitted with a more severe bit than he actually needs), he will be reluctant to go into his bridle. Conversely, if he is under-bitted he may become heavy in the hands, leaning on the reins and falling on his forehand. Introducing a horse to a curb bit must be done gradually with sensitive riding, allowing the horse to relax in his jaw and accept the bit, at the same time going forward with flexion in his neck and poll. While forward movement is encouraged by the rider's legs care must be taken that the horse does not become overbent and lean on his forehand.

Action and fitting

Bits should be adjusted so as not to wrinkle the corners of the mouth. There should be no more than about ¼ inch (5 mm) play on either side to prevent the bit sliding from side to side. Over-large bits will encourage horses to get their tongues over the bit or to become one-sided. Likewise, badly fitted bits, particularly snaffles, will cause the joint to go up against the palate which is painful and causes the horse to open its mouth. Incorrectly fitted snaffles can also pinch the bars of the mouth.

It is of the utmost importance that bits are carefully fitted to each individual. This is something that is all too often overlooked.

A well-fitted bit helps to keep the horse balanced and collected by the lever action of the bit, which should induce the horse to bend at the poll and relax his lower jaw into the rider's hand.

Mouthpieces that are thin are always more severe on the horse's mouth. Some bits are rough or twisted on one side and the smooth side should always go next to the horse's tongue.

Snaffle bridles do not offer the same degree of control as the curb bit. The action of the snaffle is to raise the horse's head, in contrast to the curb bit which should relax the jaw and flex the poll. The mildest type of bit is an unjointed snaffle.

The bars of the mouth can also be damaged by ill-fitting bits and if these areas are allowed to become calloused and chafed the horse will soon become resistant to the bit. Mouth sores are slow and difficult to heal and can result in the horse being out of work for a while because he is unable to wear a bit, so it makes sense to guard against them.

The position and effect of bits can, to a

certain degree, be influenced by the fitting of the noseband and martingale. Horses who develop the habit of getting their tongue over the bit are a problem and it is difficult to prevent them doing this other than to fit a tongue grid or to tie up the bit to the noseband. Even then, some determined horses will find a way round it. Drop nosebands help to keep the horse's mouth closed. There is a selection of bits with a high port which will also act as a deterrent. Prevention, as ever, is better than cure and the horse must be discouraged from getting his tongue over by correct bitting when he is at the breaking stage or being shown in-hand.

Materials used in the manufacture of bits
Bits are made of various metals, the best quality ones being of stainless steel. The advantages of stainless steel are that it is resistant to breakage, durable, does not corrode and is easy to clean. The steel used in the past, which had to be burnished and polished after every use, is no longer popular because of its tendency to rust. Never be tempted to buy nickel bits even though they are half the price of stainless steel ones. Nickel is a soft metal which is liable to bend and break easily. It is a false economy to buy nickel, because of the safety aspect.

Other materials used for the mouthpieces of bits are vulcanite, rubber, wood, leather and nylon.

Types of bit

Bits are generally divided into four categories. First and most popular is the snaffle, followed by the curb, then the pelham and the various gadget-type bits.

Snaffles

The most common snaffle in use is the smooth single-jointed type with loose rings. The twisted snaffle is not used as much today as it was; the thickness and severity of the mouthpiece varies but it is still a harsh bit for horses who pull. The cheeked snaffles most widely used are the Dr Bristol with a plate in the centre, the D-ring snaffle, the double-ringed or Wilson snaffle and the scurrier or Cornish snaffle. The scurrier is a severe snaffle often used for hunting and show jumping on horses with strong mouths. The D-ring snaffle, with either a metal or rubber mouthpiece, is often used for racing. The straight-bar snaffle, half-moon bar snaffle and mullen-mouthed snaffle are useful mild bits. The rubber snaffle, a very mild bit for a horse with a soft mouth, consists of a broad rubber bar with metal rings at each end. The roller snaffle which has a number of loose metal rollers round the mouthpiece is good for a horse with a hard mouth and nowadays is used a lot out hunting.

Gag snaffles should always be used with two reins, one attached normally and the other fixed to the pulley ring on the cheek pieces. When the gag rein is pulled the snaffle is forced up against the corners of the mouth, the result being that the horse should draw his head up. It is a useful bit for very strong horses or those that bear their heads down. Gags can have mouthpieces made of rubber, which is the kindest, or twisted steel, which is the most severe. It is also possible to have a very small gag bit fitted into a double bridle instead of a snaffle. This is often referred to as a groom's gag.

The action of the snaffle is on the bars of the horse's mouth, using pressure rather than a leverage action. They have rings which can be either fixed or loose. Mouthpieces can vary widely from a mild, thick one to a thin, more severe type. The straight bars and mullen types are particularly kind and therefore suitable for young horses and those with sensitive mouths. Some snaffles, if used incorrectly, have a nutcracker effect which can pinch and be very sharp on the bars and tongue. When fitting a snaffle the bit should be high enough in the mouth to slightly wrinkle the corners without making folds in the skin. A loose-ring snaffle will tend to pinch the corners of the mouth, whereas an eggbutt type will not. Large-ringed snaffles are popular for racing as they will not pull through the horse's mouth. Snaffles also come with a variety of long cheek pieces, which are useful in keeping the horse's

Eggbutt mullen mouth snaffle in metal.

Eggbutt mullen snaffle (vulcanite).

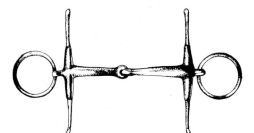

Australian loose-ring or Fulmer snaffle.

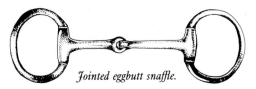

Jointed eggbutt snaffle.

Barmouth breaking bit with keys.

Wire-ring tapered race snaffle.

A fat mouthpiece.

New Zealand full spoon-cheek snaffle.

Wire-ring jointed snaffle with rollers round the mouth.

Eggbutt Dr. Bristol race snaffle.

Dee-cheek race snaffle.

Cheltenham gag.

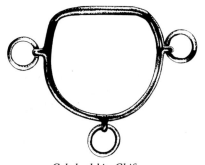

Colt lead bit, Chifney.

Short-cheek slide-mouth Weymouth and bridoon

Kimblewick, Cambridge-mouth; true Kimblewick.

Fixed-cheek Weymouth and bridoon.

Port-mouth tongue Pelham.

Rubber mullen-mouth Pelham.

Single-link curb chain.

Jointed Pelham.

Double-link curb chain.

Leather curb chain.

Elastic curb chain.

101

SNAFFLE BRIDLE.

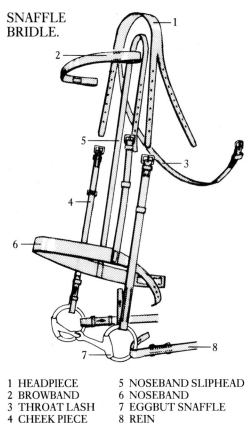

1 HEADPIECE
2 BROWBAND
3 THROAT LASH
4 CHEEK PIECE
5 NOSEBAND SLIPHEAD
6 NOSEBAND
7 EGGBUT SNAFFLE
8 REIN

head straight. Breaking bits also fall into this category, from the straight-bar type, which can have keys added to encourage a horse to salivate, to others which have a long cheek piece or a broken mouthpiece and are often used when a young horse is first ridden away after lungeing or long-reining.

Curb bits and the double bridle

The double bridle consists of a combination of a curb bit with a bridoon (snaffle). It is not severe if the cheek pieces of the curb are of medium length. The most usual pattern is the Weymouth, or the Banbury which is more severe and has a sliding mouthpiece. Double bridles are used after a good sound foundation of the mouth has been made with the snaffle.

The curb bit of the double bridle is known as the Weymouth and comes with or without a port. The Weymouth cheek can either be long or short, fixed or sliding. In the past many

types of curbs were used but nowadays it is mostly the short-cheeked varieties, either fixed or sliding. The very short-cheeked bit known as a Tom Thumb is the mildest.

The double bridle employs a snaffle (bridoon) and a curb bit in the mouth together.

DOUBLE BRIDLE

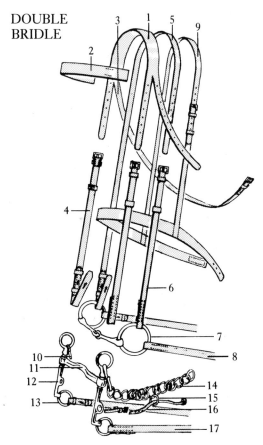

1 HEADPIECE
2 BROWBAND
3 THROAT LASH
4 CHEEK PIECE
5 BRIDOON SLIPHEAD
6 BRIDOON CHEEK PIECE
7 LOOSE-RING JOINTED BRIDOON BIT
8 BRIDOON REIN
9 NOSEBAND SLIPHEAD
10 PLAIN CURB HOOK
11 LOW PORT FIXED CHEEK CURB BIT
12 LIP STRAP RING
13 CURB REIN RING
14 SINGLE LINK CURB CHAIN
15 FLY LINK
16 LIP STRAP
17 CURB REIN

The two bits are entirely separate but are used simultaneously to effect a combined action. The bridoon is used to raise the head while the curb is employed as a means of achieving flexion at the poll. The degree of success of these objectives depends on the experience of the rider and his use of the double bridle. Perhaps the most popular of the double bridle combinations is a short-cheek curb with a loose-ring bridoon. One double which is becoming increasingly used today is the fixed-ring Weymouth with German mouthpieces combined with an eggbutt bridoon. The action of this bridle is more positive than some and prevents the horse playing with the bit. It is commonly used for dressage and in the show ring. The curb bit acts entirely on the top of the bars. The port allows space for the tongue. With a low port some of the pressure is taken on the tongue; with a high port there is no pressure on the tongue but as the bit tilts the port may touch the roof of the horse's mouth. The Weymouth should be fitted in exactly the same way as the snaffle, lying about 1 in (25 mm) below the bridoon at such a height that the curb chain remains in the curb groove when the reins are engaged. The length of the curb chain should be sufficient to allow the cheek of the bit to turn at an angle of 45° to the line of the mouth. The curb chain must lie flat. If the horse resents the metal chain it may help to use a leather or elastic one or to cover the curb chain with rubber or sheepskin.

The pelham

The pelham has become a popular bit and like the snaffle has a variety of mouthpieces. It can be used in conjunction with leather 'D's for use with single or double reins. Pelhams are a combination of snaffle and curb in one mouthpiece. The bit has rings on the ends to take a snaffle rein and also cheeks with a hook for the curb chain at the top and a ring for the curb rein at the bottom. The difficulty in using a pelham is that the action of the snaffle rein raises the bit in the mouth. This brings the curb chain out of the groove and if the curb rein is pulled the chain is likely to rub the jaw. The rider must therefore be particularly care-

ful to effect a balanced use of the two reins. A pelham with a thick vulcanite mouthpiece is a very useful bit. It is mild, yet the thick mouthpiece acts on a large area of the bars which means that if a part of the bars becomes calloused the horse will still be sensitive and respond to the bit.

The pelham should be fitted so that the curb chain lies in the groove when the bottom rein is employed. A common fault with curbs on pelhams is that they often pinch and chafe the sides of the horse's mouth and lips but this can be avoided if the curb is passed onto the outside of the snaffle ring rather than in the normal position. Pelhams should not be used on young horses as a short cut to using a double bridle. This can be seen in the show ring and tends to encourage a faulty head carriage once the horse has learned to lower it and tuck his chin in. This is a habit also to be avoided when using a double because it is difficult to overcome and the horse gets used to going behind the bridle. For a horse with a difficult mouth and those who tend to pull, the vulcanite pelham is a worthwhile aid. The many types of pelham bits are often used on the polo field.

Belonging to this group of bits is one called the Kimblewick. It was introduced in this country by Phil Oliver and named after the village in which he lives. Originally it only had a straight-bar steel mouthpiece with a curb and a large 'D' made into the bit. Nowadays it is possible to buy Kimblewicks made with different mouthpieces. With its single rein action it was originally used for show jumping but now it is more widely used, particularly by children who need something more effective than a snaffle on a strong pony.

Gadgets and bitless bridles

There are some horses whose mouths have become so hard and calloused that they need a special bit. Among these is the Army Reversible which is a curb bit, often found satisfactory on a hard puller. The 'three-in-one' bridle is attached to the rings of the bit thus there is no displacement of the mouthpiece before the curb chain action begins.

The Liverpool has a severe curb action and is often used as a driving bit.

A point to remember with severe bits is that if the rider continually hangs on to the reins this will lead to an even harder mouth, devoid of any feeling. Many pullers will go better in a non-curb bit. There is often a tendency for people to use bits which are more severe than is necessary, particularly when jumping where it will only lead to the horse refusing.

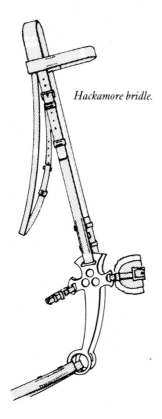

Hackamore bridle.

The Hackamore bitless bridle consists of a snaffle bridle which has no mouthpiece but the cheek pieces of the bit act as levers to bring pressure on the nose, poll and chin groove. There are many different variations on the market.

Headcollars

Nothing looks better on a horse than a good quality leather headcollar with brass fittings. Various types and fittings are available. Some prefer headcollars with a browband, which can be plain leather, white canvas or buckskin. Some headcollars are fully adjustable with buckles either side of the headpiece plus a buckle on the noseband and throat lash, enabling them to fit correctly. Foal slips and yearlings' headcollars also vary in type.

The most popular headcollars for everyday use are made of nylon and webbing. They are easy to maintain and very strong but, like leather, the quality varies considerably.

The cavesson headcollar

The cavesson is used for lungeing and consists of a very strong type of headcollar with a padded noseband, usually with metal reinforcement for greater control. It has a ring on the noseband to which the lunge rein can be fixed. They also have side 'D's to which side reins can be attached. Care must always be taken during lungeing that the cavesson straps do not pull into the offside eye.

Lungeing cavesson on top of snaffle bridle.

Bridles

Bridles should always be made of the best quality leather. Bits can either be sewn on permanently or billet and studded. The bridle consists of a headpiece, browband, cheek straps, noseband, bit and reins. Attached to the headpiece is the throat lash and when fitted there should be a space of at least three fingers between the lash and the throat. The browband should allow enough room for the ears to move. The reins can be rubber-covered, plaited, plain or laced. Some are now made of webbing with narrow leather 'stops' sewn on. Reins should always be of an average width, i.e. not too narrow. They vary from $5/8$–1 in (15–25 mm) in width and from 4–5 ft (1.2–1.5 m) long.

Correctly fitted snaffle bridle.

Plain leather rein | Rubber rein | Plaited rein | German web rein | Buckle rein | fastener | Billet fastener | Stitched rein

Laced leather reins.

Running or draw reins

These are used in the schooling of difficult horses and should always be applied with great care as they tend to make horses overbend. They can be attached to the girth, either between the horse's legs or at the sides. From here they pass up through the bit rings and back to the rider's hands, providing leverage.

Chambon

This is a schooling device of French origin which consists of a strap from the girth passing between the fore legs and dividing at the breast into two cord attachments. These pass through rings on the poll pad and are then connected to the bit. The object is to lower the head and round the back, engaging the hind-quarters.

Nosebands

The Cavesson noseband is the simplest and kindest noseband which fits high around the nose. It should not be fitted tight like a drop noseband. A standing martingale is usually attached to a cavesson noseband.

Drop nosebands can vary slightly in design but their principle action remains the same, i.e. to gain greater control. They are often employed to prevent horses getting their tongues over the bit. They should never be fitted too low as this will constrict breathing. They are most successfully used with a snaffle bridle.

Sheepskin noseband.

Drop noseband.

Kineton Noseband. A severe method of restraining a horse which pulls hard. Unlike other drop nosebands it does not close the mouth.

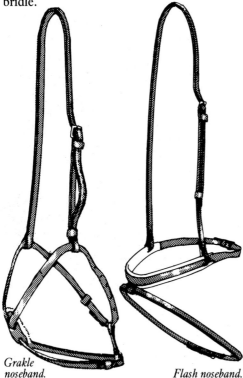

Grakle noseband.

Flash noseband.

The Grakle consists of two straps crossing over the nose, one over the bit and one under it. They are connected underneath.

Flash nosebands consist of an ordinary cavesson noseband with two straps sewn on the centre which fasten underneath the bit. They are often used in conjunction with a standing martingale.

Sheepskin nosebands help to prevent the horse from carrying his head too high and are mainly used in racing.

Browbands

Browbands should always be of plain leather; exceptions are the coloured ones used on racing bridles, show ponies and show hacks. Coloured browbands should never be seen in the hunting field.

Boots

Boots are designed for the following purposes: as protection from injury caused by jumping; brushing; over-reaching and speedy cutting; as support for the tendons, and as protection for the legs during travel.

The main types are:

Shin boots – used for jumping to protect both the fore and hind legs.

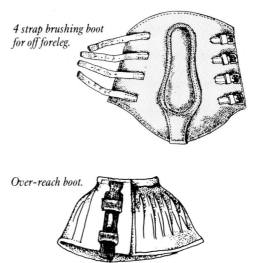

4 strap brushing boot for off foreleg.

Over-reach boot.

Over-reach boots – fitted to the front feet to guard against the hind toe injuring the front heel.

Tendon boots – designed with a strong pad shaped to the leg to support weak tendons.

Hock boots – used for travelling, to protect against capped hocks.

Travelling boots – protect the whole of the lower leg from the knee or hock to the coronet

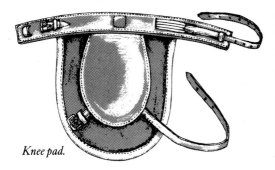

Knee pad.

Knee-pads – for travelling and for protection against stumbling on the road. The skeleton knee-pad is more practical for riding.

Coronet boots – mostly used during polo matches for protecting the coronet from injury by the other hooves.

Brushing boots – protect the tendon and inside of the fetlock joint from contact with the opposite leg and can be used in front and behind.

Speedicut boots – similar to brushing boots, for the fore leg, fitted slightly higher to protect against speedicuts.

Heel boots protect the point of the fetlock which may come into contact with the ground during fast work and jumping.

Travellers fitted over travelling bandages for double protection.

107

and combine knee-pad/hock boot and padding. Used during travel, they are made of various materials including plastic, wool, sponge and nylon with buckles or Velcro fastenings. They have the advantage over the combination of bandages and hock boots/knee-pads in that they are not fastened around the joint and are therefore more comfortable for the horse. They can also be used in the stable to protect the horse from capping his hock. Some types of 'travellers' need stable bandages fitted underneath them to prevent them from slipping.

Pricker boots – used mainly on young racehorses. They are studded with tacks and fitted over a bandage to prevent the horse tearing the latter off.

Poultice boot – for holding poultices in place (see also page 108).

Sausage boot – usually fitted to one leg as a buffer to protect the elbow when the horse lies down, thus preventing a capped elbow.

Kicking boots – fitted to the hind feet of mares for the purpose of protecting the stallion from being kicked during covering.

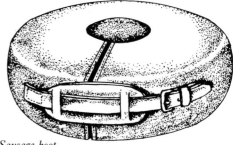

Sausage boot.

Yorkshire boot – a rectangular piece of (usually) woollen material with a tape or Velcro around the centre. They are tied just above the hind fetlock and folded over to give double protection from the opposite leg.

Fetlock ring boot.

Fetlock ring boot – a hollow rubber ring fitted over the fetlock to protect against brushing.

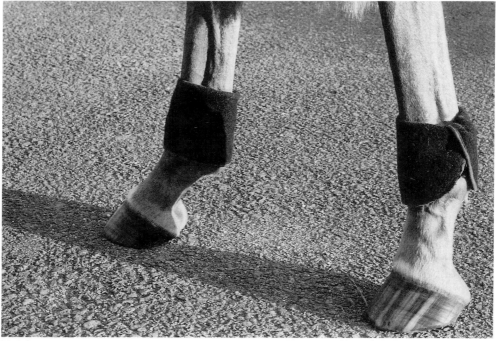

Yorkshire boots.

Fitting boots

Fitting boots of any sort is critical because bad fitting can cause permanent damage to the tendons and joints. A boot which is fitted too loose is just as dangerous as one which is too tight because as it comes loose the horse could tread on it and fall. It will also allow stones and

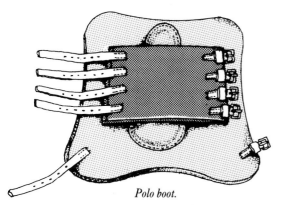

Polo boot.

mud to get between the boot and the leg causing chafing. A tight boot can mark the leg and cause long-term damage to the tendon and the joint. It is important to have an even tension, whether the boots are fastened by buckle, Velcro or elastic. For this reason it is good practice to have the same person fit all the boots on one horse and check for even tension. All boots, especially leather ones, should be well maintained, washed and dried thoroughly after each use; mud and water will wear them badly.

Brushing boots should fit well under the knee and down to the fetlock joint in order to protect the whole of the cannon bone and tendon. These boots are recommended for everyday use as a safeguard against the horse knocking himself with the opposite leg. This applies particularly to young, unfit and weak horses. Boots which become muddy and wet can set up an irritation and sometimes cause swelling, which must be recognised as such and not mistaken for a sprained or damaged tendon. Most brushing boots are lined with rubber or felt and made of leather, plastic or

cloth, with Velcro or buckles for fastening. Front brushing boots have either one-, two-, three-, or four-strap fastenings, while hind boots generally have five straps. They can be bought either off the shelf or made to measure. Brushing boots or Yorkshire boots can be fitted to prevent concussion of the fetlock joint which can if not treated form a callous and enlargement.

Good quality hock boots are made of leather and wool with a leather strap top and bottom. It is vital when fitting them to ensure that the bottom strap is loose enough to allow the horse to flex his joint normally as he walks. Hock boots are costly and no matter how well they fit always seem to irritate the horse causing him to fidget and kick. Some animals take such a dislike to them that they cause more trouble than they save. 'Travellers' are often preferred to hock boots for this reason.

Bandages

Bandages are used for protection, support, warmth and veterinary reasons, the exception being the tail bandage whose effect is purely cosmetic. Bandages are made of wool, flannel, stockinette, crepe, cotton or a combination of wool and stockinette in the case of Sandown bandages. Most bandages need some padding underneath to protect the leg against direct pressure from the bandage itself. All bandages should be fitted with an even pressure, prefer- ably by the same person on one horse. The surface needs to be kept flat and free from creases to avoid ridges forming on the horse's skin. Be sure that the tapes are never tighter than the bandage. Change bandages at least every twelve hours. Use plenty of under- padding where there is an injury. The tapes or Velcro must be kept flat and the ends of the tape tucked in to discourage the horse from chewing.

Stable bandages are made of either wool or flannel about 7–8 ft (2.1–2.5 m) long and 4–5 ins (10–12 cm) wide. They should be fitted from just below the knee or hock to the top of the coronet, but not tightly. Their

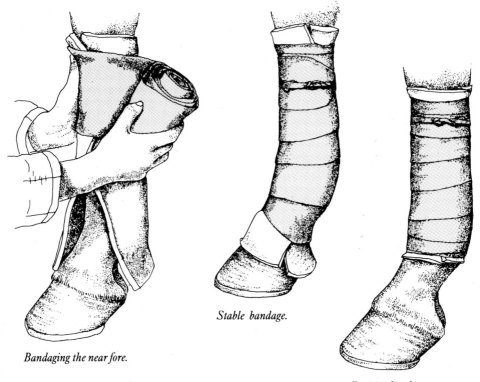

Bandaging the near fore.

Stable bandage.

Exercise bandage.

purpose is to provide support, comfort and warmth after work for at least two or three hours, but preferably overnight. Used this way they will prevent filled legs and windgalls developing. With hay or straw underneath they will help to dry off wet legs. Stable bandages can be used as cold-water bandages as a preventive treatment, after being thoroughly soaked in cold water. They must be changed frequently as the warmth of the leg will eventually dry them out and affect the tension.

Thermal bandages,which are used in the stable for warmth, have to be fitted directly to the leg, without any padding underneath, if they are to be effective in promoting the horse's circulation.

Sandown bandages are used in the stable, for travel and for exercise. The woollen end is fitted against the leg with the stockinette securing the bandage on the outside. Like stable bandages they are about 5 ins (12 cm)

wide. As with thermal bandages they can be used without under-protection.

Exercise or pressure bandages are used for the horse to work in to support the tendons, especially important for weak or strained tendons which need this extra support and protection from injury. The exercise bandage is made of crepe, elastic or stockinette about 2–3 ins (5–7 cm) wide. Gamgee or some other similar under-padding must be fitted and the bandages applied from just below the knee to above the fetlock. It is unwise to use these bandages unless absolutely necessary because the tension is critical and harm and possibly permanent damage can be caused through ill-fitting. If they become loose they can trip up a horse and cause him to fall. Badly fitted bandages can cause a horse to break down so if you are unsure or inexperienced it is wise to find someone experienced to do the bandaging.

Newmarket bandages are made of cotton or

crepe with some stretch and make ideal tail bandages although they can cause scabbing if fitted immediately after the tail has been pulled. They are more secure than stockinette or exercise bandages but should be removed once they become wet to avoid injuring the horse when they dry out. Guard against fitting too tightly or for long periods.

Bandage under-padding

Apart from thermal and Sandown bandages they must be fitted on top of a layer of material to protect the leg from excessive or uneven pressure of the banadage. It should be cut large enough to extend above and below the bandage to avoid the bandage cutting into the leg at either end. There is a variety of bandage padding on the market and although price might be influential in the selection it is a false economy to buy a cheap material. Gamgee, although relatively expensive and having a short life, is one of the best materials. It can have its edges sewn once it has been cut to shape to prevent fraying and distortion through use. When it becomes worn and thin in places it should be discarded as it will no longer afford even protection. Cotton wool can be used for veterinary purposes, providng a good thickness is allowed. It will, however, wear and tear after little use. It is not recommended for use with stable or exercise bandages.

Of the ready-made paddings available Fybagee is amongst the most popular as it is very hard-wearing. It is available in various sizes to fit any part of the legs, for use in the stable, on exercise and during travel. Its great advantage is that it is machine-washable.

Clothing

The principal reasons for clothing a horse are to keep him warm and to save his body expending more energy than necessary to do this, thereby not making the most efficient use of his food. Clothing will improve the appearance of the coat and help maintain it in good condition as well as keeping the horse warm and dry. The circulation will be improved with less likelihood of the horse catching a chill.

The various uses for rugging are in the stable, on exercise, travelling and while grazing. At night a stabled horse may use individually or in combination a cotton sheet, a woollen blanket, a night rug and stable bandages, depending on the time of year, temperature, environment, condition of the horse, his age and his work routine.

At other times clothing is needed for appearance; for example, when parading at the races or waiting for a competition to start. Clothing can be used to make an overweight animal sweat by putting a polythene plastic sheet under the rugs. Rugs always need fitting very carefully otherwise it will encourage a horse to chew and tear at it. A clothes bib will prevent a horse from doing this.

Rugs of an inferior quality should be avoided because they will not fit well; they are usually shallower and less durable. Straps, buckles and stitching wear more quickly too, making the rug unsafe. Surcingles, fillet strings and straps are often extras to be added to the basic price of the rug. Nylon straps are becoming more popular now and are very practical.

A rug should be large enough in the chest and fit the horse for length. Night rugs need to be slightly larger than day rugs and sheets. Day rugs and sheets may be slightly shorter in length. If the rug is too short to allow several inches gap at the chest it will slip back and rub the shoulders and put undue pressure on the withers. Horses with high withers often need extra protection in the form of a piece of foam, sheepskin or woollen blanket or a wither pad. Occasionally an odd-shaped horse will need a rug made to measure.

Some rugs today are fitted with two surcingles, one over the withers and one over the loins. The latter must be kept slack so as not to aggravate the horse. There should not be any wrinkles under the front strap otherwise it may cause rubbing and chafing.

The American fastenings of cross surcingles hold the rug in place, avoid pressure on the horse's back and eliminate the need for a roller. These are particularly useful for narrow horses.

Types of rug

Quilted rugs, such as Polywarms and the various other types now on the market, can often replace both a blanket and rug. They are filled with insulating material and lined with either synthetic or natural fibres which makes them very warm and light and easy to clean.

Jute rugs make excellent everyday night rugs and are relatively inexpensive. They are often called dealer's rugs as they are frequently used on horses at sales.

Flax night rugs are excellent. They wash well and are long lasting, but are twice the price of jute rugs.

Day rugs are made of woollen material and can be of any colour, usually with a contrasting binding.

Summer sheets can be used not only for keeping the dust and flies off the stabled horses in summer but also as an undersheet to keep dust and grease out of other clothing. Occasionally a thin-skinned horse will object to wearing a woollen blanket next to his skin and if so, the fitting of an undersheet will help. Sheets are made of cotton or linen and should always be washed at least once a week. Horses that are excitable when travelling will benefit from only wearing a sheet.

Blankets are traditionally made of wool and referred to as Witney blankets. They usually have yellow, black and red stripes which normally run horizontally when the blanket is on the horse, but in cases where the blanket hangs below the rug it is more practical to fit the blanket with the stripes across the back. Some blankets can be made into day rugs, which are both warm and attractive. Always ensure that the blanket fits well under the rug, is folded in front to stop it slipping back and held in place with a roller or surcingle.

Sweat rugs and towelling sheets are useful for cooling horses after fast work or washing. Many people use them for travelling.

Exercise rugs and sheets, often called quarter sheets, are made of various materials depending on the time of the year and weather. Woollen-lined waterproofs and plain woollen ones are the most popular. Matching hoods can also be purchased for exercising in the winter. Competition horses often wear hoods and full-size waterproofs while awaiting competitions in wet weather.

New Zealand rug. The design and quality of New Zealand rugs can vary as much as with any other rug but because the horse is left to run free with a New Zealand rug it must fit correctly otherwise it can be dangerous. A well-designed New Zealand rug, with leg straps in the right place, will stay in place after the horse has rolled. Inferior makes are less waterproof, need more drying and do not fit as well, which makes them unsafe to leave on a horse for any length of time. With the horse having to exercise in a New Zealand rug it is imperative that it fits properly on the shoulders and back and does not rub against the withers or girth area. If it is a tight fit when first put on, it will soon chafe the horse as he moves. A piece of sheepskin sewn under the withers and shoulders of the rug will protect the horse to a certain degree. It is prudent to have a replacement New Zealand rug in case of damage or to allow a wet rug to dry off before being used again. Poor quality straps and buckles will wear badly and soon become unsafe as they rust and tear. A broken strap or buckle will cause the rug to slip and can be extremely dangerous. A horse with a slipped New Zealand rug which becomes loose around his legs may panic and gallop, trying to kick his way out, sometimes with disastrous results.

Further information about New Zealand rugs can be found on page 156.

Rollers

Rollers should always be fitted with a pad underneath to prevent rubbing the back and withers sore. This can be made of foam, sheepskin, felt or similar material. An arch roller must be wide enough so as not to pinch the horse's back and be padded sufficiently to prevent irritation to the back. Badly fitting rollers are often the cause of white hairs growing in that area. A breastplate fitted to the roller will mean the roller will not have to be done up so tightly and also prevent it slipping back. Rollers made of webbing invariably wrinkle with age and can then cause galls to form under the girth area. Leather rollers must be preserved by regular saddle soaping or oiling to prevent them going hard and cracking. Double-strap rollers of either leather or webbing are preferred to the single-strap ones, which tend to cut into the horse. Many surcingles which come attached to rugs are not sufficient to keep a rug in place, especially if a blanket is fitted as well. Rugs and rollers with inferior strappings can damage horses' backs so they should be regarded as a false economy.

Preparing a horse for travel

The various factors to consider when travelling a horse are: the time of year; weather conditions; the length of the journey; the horse's age and temperament; how many horses are travelling together; and the type of vehicle. Some form of leg protection will be needed, together with rugs, tail bandage and tail guard. If stable bandages are used they must be fitted with an under-padding of gamgee or similar material. Knee-pads and hock boots may need a layer of cotton wool under the top strap for long journeys to give extra protection against rubbing. Where leg protectors or travelling boots are used they will combine the role of both knee-pads or hock boots with bandages. This is often preferred as they are simple to use, easy to keep in place and do not cause the aggravation which hock boots often do. These travelling boots and leg protectors are usually made of rubber or plastic with Velcro or leather fittings; they must fit the horse well or they will slip and twist. This can also cause rubbing, which is particularly troublesome if a tendon is marked. Conversely they must not be fitted too tightly especially for long journeys, or they will do a lot of harm. More bandages are now fitted with Velcro instead of tapes and are therefore easier to fit. Bandage tapes can often be fitted too tightly and mark the tendons.

It is often practical to fit brushing boots and over-reach boots in front for short journeys when the horse is tacked up, such as for hunting, but this is not recommended for longer journeys because of the lack of protection to the tendons and joints. However, over-reach boots can be used in conjunction with bandages or 'travellers' for longer journeys to guard against the horse treading on himself. It is common practice to fit only front boots and knee-pads to racehorses who, like many fit horses will often kick out during travel. They risk injuring themselves if bandages or leg protectors slip. The other exceptions where horses are better without any leg protection, are youngstock and nervous horses who may panic with something on their legs. Most horse boxes are fitted out with rubber or coconut matting to help protect the horse against injury.

For short journeys a tail bandage may be sufficient if the horse travels well but for those who tend to sit on their tail and for long journeys a tail guard should be used as well. If in hot weather no rugs are needed, a roller or surcingle with a wither pad will have to be fitted to attach the tail guard.

If a leather tail guard is used it must be supple or it will rub the underside of the dock. A lightweight roller should be chosen in preference to a heavy leather or anti-cast type which will cause the horse to sweat. A breast-plate will eliminate the need to do the roller up too tightly and still be safe from slipping. A poll guard will be necessary for very long journeys or where a horse is prone to throwing up his head. It should always be used for air travel because there is normally little head-room on the aircraft.

The amount and type of clothing will vary according to the above-mentioned factors. It can be anything from a single sweat sheet in the summer to a combination of rugs for a long journey in the winter. Travelling rugs specially fitted with a surcingle are ideal and can be used over a sweat sheet or lighter rugs. If a day rug is used and turned back to avoid the horse sweating on his shoulders and neck, care must be taken to see that it does not slip back under the roller. Very often a sweat sheet with a woollen rug on top will be adequate. However, on a long journey it is practical to fit first a sweat sheet then a towelling, cotton or linen sheet on top of which a woollen rug and/or blanket or travelling rug can be added. This allows for the top layers to be removed or added depending on the horse's requirements during the journey.

The type of vehicle will also play a part in deciding what rugs to use. For example, trailers and cattle trucks are particularly cold whereas custom-made horse boxes, especially the modern luxury types, often have air-conditioning. The number of horses travelling together will affect the temperature inside the vehicle.

It is very easy for a horse to catch a chill or sweat profusely and lose condition because of inappropriate clothing. This is always serious and can damage a horse's fitness. If ever a colt has to travel with fillies or mares he should not be able to touch them. He may fret and sweat so he must be watched carefully to ensure that he does not catch a chill.

Martingales

The main types of martingale are the standing, running, bib, Irish and Market Harborough. **Standing martingales.** The purpose of the standing martingale is to encourage the horse to go with his head in the correct position, level with his withers. It should be fitted so that it provides adequate freedom, particularly over fences, and only comes into effect when the horse tries to raise his head too high. In the case of an awkward horse, such as one who rears, it can be shortened to prevent this. A standing martingale is in many ways kinder than a running martingale because the pressure is put on the nose and not the mouth. For this reason it is very useful for young horses who are being ridden away during the early stages. A correctly fitted standing martingale will allow the horse to jump and perform most work and it will prevent him jumping with his head in the air. There should be no slackness in the martingale between the horse's chest and the girth, in which the horse could catch his foot. A rubber ring at the junction where

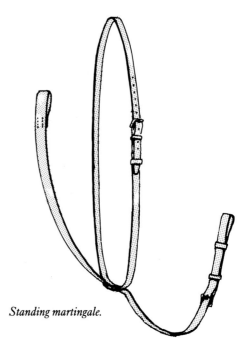

Standing martingale.

the martingale joins the neck strap will prevent this and allow for some slack between the noseband and the neck strap. The types which have a buckle at the end near the noseband are often preferred to those with the buckle near the girth as they are easier to adjust and to clean.

Running martingale. This is designed to act on the bars of the horse's mouth when he throws his head and to steady him when he jumps. If too tight its action can be severe and is not as kind as many people believe. If it is fitted too loosely it is totally ineffective. It must reach the top of the withers and not affect the tension on the rein between the horse's mouth and the rider's hands. The reins must be fitted with a rubber or leather stop to prevent the rings of the martingale catching on the billet studs, buckles or bit. If this is overlooked the rings can get caught and cause the horse to go backwards or rear, often with disastrous consequences. If the rider should find that the horse has caught his teeth in a martingale ring he should dismount instantly and put things right. If martingale stops are not fitted to the reins a running or bib martingale should not be used. The only exception to this are reins

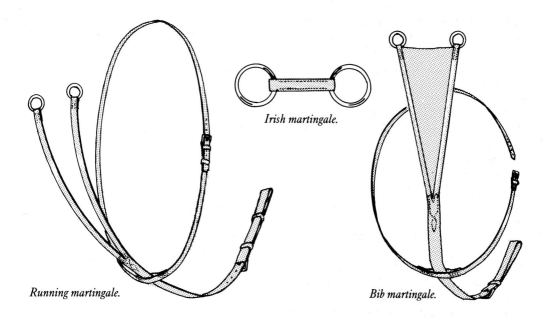

Irish martingale.

Running martingale.

Bib martingale.

with sewn ends. The ring on a curb bit is smaller than the ring of the martingale and the same thing can happen if stops are not fitted. **The bib martingale.** This is similar to the running martingale and has the same effect but it has a piece of triangular leather joining the two rings together. It is very useful for young horses who may try to play with the rings and could get their teeth hooked up.

Irish martingale. Sometimes referred to as 'rings' or 'spectacles', the Irish martingale is a piece of double leather about 4 ins (10 cm) long with a ring on each end through which the reins are passed to keep them on either side of the horse's neck. They are used on horses that throw their heads about, in particular race-horses. During a race they are considered an extra and must be included when weighing in. **Market Harborough.** This martingale has twice the length of split rein with a hook on each end which is passed through the bit and clipped onto reins which have D-rings at equal points on either side. The Market Harborough operates by putting pressure on the mouth when the horse raises his head. This action is immediately halted when he lowers his head.

Girths

The most common girths in everyday use are the Atherstone, Balding, lampwick, three-fold, web and elastic, cordstring and nylon.

Safety is of key importance so only good quality strong girths with hard-wearing buckles and stitching which is in a good state of repair, should ever be used. At the slightest sign of wear a girth should be removed from the tack room to avoid anyone accidentally using it until it is repaired. Maintenance is a crucial factor of any tack and equipment, particularly girths, which have to take such considerable weight and strain. They must therefore be kept clean and oiled to ensure that they do not become stiff and in danger of cracking or the stitching coming undone. Inferior stitching will rot with age and cheap buckles will bend, so every care should be taken to check tack regularly. String, nylon and lampwick girths will, in the course of time, start to fray, in which case they should be discarded before they cause an accident as they cannot safely be repaired at that stage. The slightest wear in a girth, whether it be of material or leather, will be inclined to cause girth galls and sores which can make the horse unrideable for some time. Where repairs have been made it is important to again satisfy yourself that they are safe and do not rub the horse because of their stiffness at first.

Lampwick girth.

Lampwick girth. This is made of tough tubular wick material which is both soft and supple and therefore ideal for thin-skinned horses. The ends are best made of rawhide leather which will outlast ordinary leather. They must be brushed thoroughly once they are dry and washed regularly. Although they are gentle on the horse's skin they are liable to fray, especially where a martingale is fitted, and are then unsafe to use.

Atherstone girth. These girths are specially shaped where they fit behind the horse's elbows to prevent chafing. This does, however, result in uneven pressure. Some Atherstone girths are fitted with an elastic insert between the leather and the buckle.

Balding girth. Designed by William Balding, this is probably as common as any, it gives the elbows room and helps to prevent galling although it does not allow for even pressure. The girth has three straps of plaited leather with the centre reinforced.

Three-fold girth. This is made from one very wide piece of leather, folded three times.

Some have a flannel inlay. The rolled edge should be fitted to face the front of the horse, as the other side which is open would rub. Because of its thickness it does cause horses to sweat under the girth more than most.

Web and elastic girth. These girths are useful for racing or jumping as the elastic insert allows for flexibility when the horse is in full stretch. If they are used with a surcingle that too must have an elastic insert.

German cordstring and nylon girths: Similar in design, both girths have three cross-weaves in the middle and one either end. The end ones are divided to each buckle. These cross-weaves should be checked daily for fraying. The cordstring type is stronger than nylon and does not cause the horse to sweat so easily.

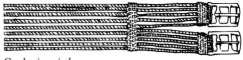

Cordstring girth.

Stirrup irons

Stirrup irons should always be made of stainless steel. It is a false economy and dangerous to use soft, cheap nickel. If a horse should fall nickel irons will buckle, perhaps trapping the rider's foot with disastrous consequences. Irons must always be large enough for the rider's boots. Choose irons that are heavy as they always hang down well. Some irons are made to curve away at the top, which adds to comfort and saves wear on riding boots.

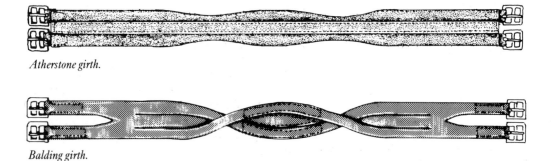

Atherstone girth.

Balding girth.

Stirrup leathers

Stirrup leathers should be of the best quality leather with stainless steel buckles, which need checking regularly for any cracks or flaws in the metal. Nickel buckles are too soft and liable to break. The buckle should be kept clean and free-moving with the buckle tongues working easily. When tack cleaning, the stirrup-leather buckle should always be undone to ensure it is working properly, the leathers checked for cracks and wear and the stitching inspected for safety. Any poor stitching should be renewed promptly and the leathers not used until this is done.

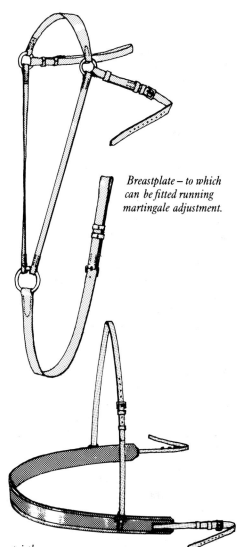

Breastplate – to which can be fitted running martingale adjustment.

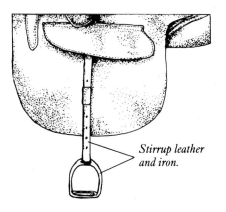

Stirrup leather and iron.

The leathers should be changed to alternate sides of the saddle to prevent uneven wear and stretching which is inevitable especially when the same person is using that pair for any length of time. Buffalo hide leather is the worst kind for stretching and will eventually become narrower at the buckle end. Rawhide stirrup leathers are much harder, less likely to stretch and often used for show ponies. Extending leathers enable disabled or short-legged riders to mount by unhooking an extension which can then be re-hooked on to the buckle end once the rider has mounted.

Breastgirth.
The breaststrap can be either leather, elastic, webbing or covered in sheepskin.

Breastplates

Leather breastplates are mainly used to prevent the saddle slipping back. They consist of a leather strap round the horse's neck which is

secured to the girths between the fore legs. The neck strap has light leather attachments that go to the 'D's in the front arch of the saddle.

A breastgirth is a wide piece of material or leather that goes from one side of the girth, round the horse's chest to the other side and has a thin leather strap over the horse's neck to hold it in place. They are mainly used in racing.

Cruppers

Cruppers are used to prevent a saddle slipping forward onto the withers. They are often required on fat ponies who don't have good shoulders or withers. They consist of a padded leather loop which is worn under the tail at the dock and is attached by a strap to the centre of the back of the saddle.

Saddles

The illustrations show the types of saddle most commonly used for the different disciplines. It is, however, entirely possible to double up the use of a saddle in some cases, for example a hunting saddle is also suitable as a show saddle. A purpose-built show saddle or dressage saddle is not, however, practical for jumping. Polo can be played in either a general-purpose or jumping saddle. Either of these saddles might also be comfortable for long-distance riding, but for the real enthusiast there is now a purpose-built long-distance riding saddle available.

SIDE SADDLE

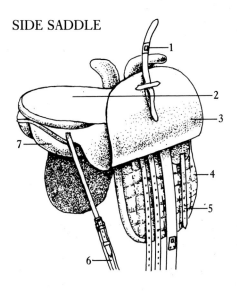

1 HOOK TO KEEP
  FLAP IN POSITION
2 SEAT
3 OFFSIDE FLAP
4 PANEL
5 QUILTED PANEL
6 POINTS STRAP
7 BALANCE STRAP

SIDE SADDLE.

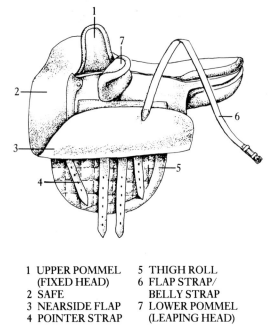

1 UPPER POMMEL
  (FIXED HEAD)
2 SAFE
3 NEARSIDE FLAP
4 POINTER STRAP
5 THIGH ROLL
6 FLAP STRAP/
  BELLY STRAP
7 LOWER POMMEL
  (LEAPING HEAD)

How to fit a saddle

Fitting a saddle is done according to the horse's conformation, the type of work he has to do and the size and shape of the rider. Ideally each horse would have a saddle made to measure but for most owners this is out of reach both financially and for practical reasons. The fitting of a saddle is a specialist job because the horse's conformation must be carefully studied. Any good saddler will be able to fit a saddle and advise the owner accordingly. For example, a horse with high, lean withers will not need the same type of fitting as one with a wide back and flat withers. A metal coat-hanger bent to the shape of the horse's withers will give a guide as to the angle needed for the saddle and this can be given to the saddler for a pattern. The following points must be considered when fitting a saddle: (a) there must be no pressure on the top of the withers caused by the front arch; (b) the horse's spine must be free from any pressure

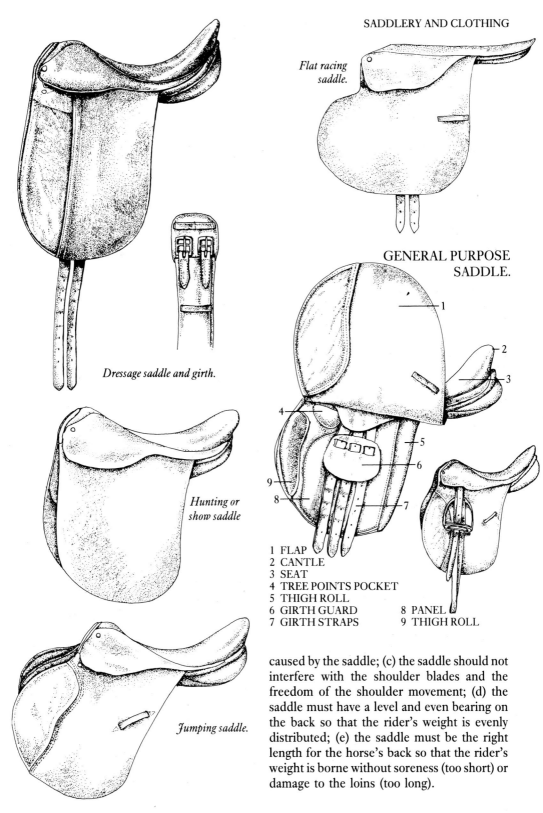

Flat racing saddle.

Dressage saddle and girth.

Hunting or show saddle

Jumping saddle.

GENERAL PURPOSE SADDLE.

1 FLAP
2 CANTLE
3 SEAT
4 TREE POINTS POCKET
5 THIGH ROLL
6 GIRTH GUARD
7 GIRTH STRAPS
8 PANEL
9 THIGH ROLL

caused by the saddle; (c) the saddle should not interfere with the shoulder blades and the freedom of the shoulder movement; (d) the saddle must have a level and even bearing on the back so that the rider's weight is evenly distributed; (e) the saddle must be the right length for the horse's back so that the rider's weight is borne without soreness (too short) or damage to the loins (too long).

119

To determine the correct fitting it is best to have someone sit in the saddle as the weight of the body will indicate the pressure it will place on the horse. Undue padding of the saddle flaps will prevent the rider from having a close contact through his legs, with the horse. Friction from the saddle rocking will cause a sore back. Horses with excessively high or flat withers or any abnormalities on the back will require a specially fitted saddle. Do not be tempted to use any saddle for convenience as it may cause permanent damage.

A new saddle will take time to settle into the horse's back and become supple. The rider's weight, particularly after a long time, will make the saddle more supple as it becomes warmer from the horse's back. At all times there should be at least 1 in (25 mm) clearance from the saddle arch. A saddle that sits too high will look ridiculous and be uncomfortable for the horse so it should be looked at again to ensure that it is a correct fit. New saddles invariably look as if they are sitting on top rather than into the horse but they will settle with use.

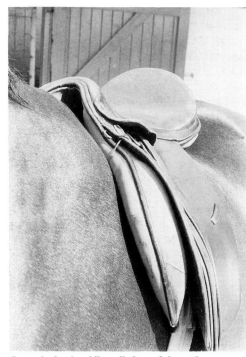

Correctly fitted saddle well clear of the withers.

Numnahs and saddle cloths

It is now common practice to use a numnah although some traditionalists frown upon them. They do, however, serve to relieve the pressure on the horse's back, especially at a time when the horse's back is soft when he first comes into work. The disadvantage of numnahs is that because they fit so close to the horse's back they restrict the flow of air along the spine. They can also slip and become wrinkled under the saddle which will pinch the skin and be very uncomfortable for the horse. Care must be taken to see that they are fitted correctly at the outset.

The choice of numnahs and saddle cloths is wide and will depend on personal preference. Invariably, higher quality ones are of better design and fitting; inferior makes will be less durable and not stay in place so well. Most types and makes are machine-washable, which is practical, but thick sheepskin numnahs are bulky and take more drying. This latter type keeps the saddle off the horse to the extent that the rider will not have a direct feeling from the horse's back. They are, however, useful in protecting the horse's back when he has had back trouble. Some, especially the foam-padded types, will cause the horse to sweat more than others.

Cleaning and storing saddlery and equipment

The atmosphere in which saddlery and equipment is kept is critical because damp conditions will make leather mouldy and hot conditions will make it hard and crack. A dry, warm atmosphere is ideal. Bad cleaning and maintenance will ruin leatherwork in no time. All saddlery and equipment should be checked before it is stored and any repairs carried out first.

Rugs and other clothing should be stored away in rug trunks with moth balls amongst them. From time to time rugs should be taken outdoors to air, given a good shake and checked for damage from moths or mice. The buckles on rugs should be smeared with Vaseline to protect against tarnishing during storage. All equipment not in use should be undone and taken apart as buckles will stiffen and become difficult to use. Leatherwork will need thorough cleaning and oiling before being put away.

Saddles should have the stirrup bars greased and a saddle cover or piece of material placed on them to protect against dust and damage. Do not store saddles on top of one another or on an iron saddle rack as this can easily damage the lining and spread the tree.

Bridles should be completely dismantled and either hung in the tack room or wrapped in brown paper and laid flat in a trunk. If the buckles are left done up they will tarnish and become stiff. The leather underneath them will harden and rot from the buckle. Ideally the buckles should be undone and the strap slipped through the keeper.

With due care leather will last a lifetime and the initial capital outlay will be rewarded with years of good service. A knowledge of saddlery is an essential part of good stable management because the days where one groom was responsible for the tack alone are long gone.

The day-to-day cleaning and maintenance of saddlery and equipment will ensure good service and reliability. All tack should be washed daily in warm rather than hot water, using a cloth or sponge, and then soaped with a proprietary saddle soap. The metal, i.e. the bits and stirrups, should be wiped with a soft cloth and/or polished with a metal polish. Leather must not be soaped without firstly being washed or the grease will only compound and form 'jockeys', i.e. small lumps of grease. It these have been allowed to accumulate they can be removed by rubbing with a ball of horse-hair. Badly cleaned leather will eventually harden and be very difficult to get back into good order. Wet leather must be dried gradually and not by direct heat or it will harden and crack. When it is very wet it may need oiling once or twice, then it can be left to dry naturally before being saddle-soaped. Girths and saddle linings will need special attention as grease will build up and cause

saddle and girth sores if they are not cleaned properly. Webbing girths need to be dried thoroughly before being brushed with a stiff brush to remove mud, sweat and grease.

Boot polish should not be used on most saddlery but can make a leather roller look very smart. Glycerine saddle soap is excellent for show tack as it does not darken the leather. Too much saddle soap on the top of saddles is unnecessary and will make the rider's breeches dirty. Linen or serge linings on saddles can be brushed with a stiff brush to remove hairs and sweat once they are dry. If they are very dirty they can be washed with warm water and allowed to dry in the open air if possible, but not against direct heat. Whitening is sometimes applied to the outer edge of linen-lined saddles for showing, which looks smart. Suede or reverse-hide leather saddles need only to be sponged and brushed when dry.

New tack may need darkening and this can be achieved by alternating saddle soap and oil. The quickest and most effective way is to warm a little oil and brush it on before ever applying saddle soap. This will make it soak in much more easily. Take care not to overheat the oil. Buckles may need a spot of oil from time to time to keep them supple. Stirrup rubbers should be taken out of the stirrup to be washed otherwise they will tarnish the stirrup.

Always use a leather punch for making new holes; anything else will cause the leather to tear.

Saddlery that has been allowed to get very dirty will be easier to clean if a handful of soda crystals is diluted in warm water. Whenever there is a contagious infection in the yard all equipment and saddlery belonging to the infected horse should be isolated, sterilised in warm soda water and dried thoroughly before it is ever used on other horses again after the infection is cured. Anything which has been in contact with an infected horse and is difficult to clean thoroughly should be burnt or thrown away to prevent further infection.

Tack cleaning and maintenance is such an important job that it should not be hurried or skipped.

Checking tack for wear and tear

Each time the tack is cleaned it should automatically be checked to ensure that it is in a good state of repair. Safety is of the utmost importance and any wear and tear should be dealt with promptly.

Bridles can be reinforced where the headpiece and reins join the bit as this is one point which, because the leather is folded, will more easily crack through wear. The saddler can sew a piece of leather under the billet so that the leather is doubled where it goes around the bit. This must be kept supple or it will be difficult to undo. Buckles and billets must be checked for cracks and wear. All stitching should be tight and safe with any suspect stitches renewed promptly. The metal parts of bridles were often made of steel, which had to be burnished regularly to avoid rust. This is time-consuming and less practical today so stainless steel is preferred because it is easily

maintained and will not rust. Nickel is also to be avoided as it is soft and easy to bend. Although the rubber on reins can be replaced new holes are punched when the new rubber is fitted, so for this reason it is generally only wise to do this once as the reins will thereafter become weaker and unsafe. Saddles, rollers and wither pads, indeed anything with a padded filling, will need to be re-stuffed from time to time. This should never be overlooked because the horse's back will be rubbed sore. Rubber stirrup treads will eventually wear thin and need replacing.

In general, stitching, buckles, billet studs and padding need constant attention to keep them in a safe state of repair. Prior to a competition, tack, including spare sets, will need double-checking.

Driving Harness. Harness design varies slightly depending on its use i.e. whether for single, pairs, tandem, or a team, and its purpose.

DRIVING HARNESS FOR A SINGLE HORSE.

1 BROWBAND	4 CHEEK PIECE	9 HAMES	16 BREECHING
2 HEADPIECE	5 NOSEBAND	10 COLLAR	17 DRIVING REIN
3 ROUND WINKER	6 LIVERPOOL BIT	11 GIRTH	18 CRUPPER
	7 BEARING REIN	12 BELLY BAND	19 LOIN STRAP
	8 HAME STRAP	13 TUG	20 BACKSTRAP
		14 SHAFT STRAP	21 REIN TERRET
		15 TRACE	

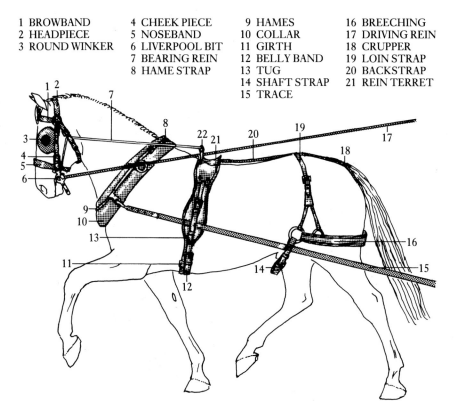

CHAPTER TWELVE

FARRIERY

General considerations

A good relationship between farrier and owner is as important as that between vet and owner because the horse's welfare is at stake. Neglecting regular attention to the horse's feet whether or not the animal is shod and in work can result in injury, illness, disease, infections and lameness, sometimes with irreversible consequences.

One of the first considerations when buying a horse has to be the economics and the owner's financial limitations will dictate how many horses will be kept and so on. But within that budget the owner must make adequate provision for regular visits from his farrier and

not 'make do' just to suit his pocket. The old adage of 'no foot, no horse' is as true now as ever and the farrier's role must never be underestimated.

The object of shoeing a horse is to enable him to work on varying surfaces without damaging his feet. The hoof is liable to split and crack as a result of concussion on hard and uneven ground and will inevitably render the animal footsore and unable to work if he is unshod. The other reason for shoeing is surgical, as a means of correcting faults in the hoof. Horses which are unshod and kept at grass will require attention in the form of trimming of the feet, at least every four to six weeks. The exception to this is young growing horses,

ANATOMY OF THE FOOT.

1 EXTENSOR PEDIS TENDON
2 LONG PASTERN BONE
3 SHORT PASTERN BONE
4 CORONARY BAND
5 INSENSITIVE LAMINAE
6 SENSITIVE LAMINAE
7 CANNON BONE
8 CRUST OF WALL

9 SENSITIVE SOLE
10 INSENSITIVE SOLE
11 LONG INFERIOR SESAMOID LIGAMENT
12 FLEYOR PERFORANS TENDON
13 NAVICULAR BONE
14 SENSITIVE FROG
15 INSENSITIVE FROG

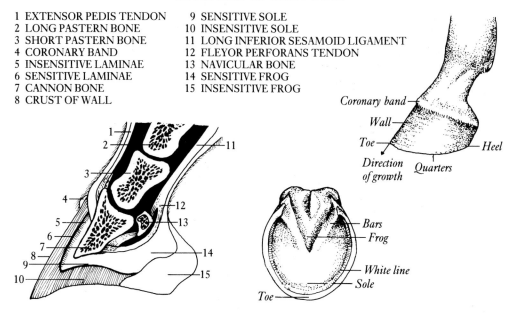

especially those between foalhood and two years who will need more frequent visits, say, at two- to four-week intervals. There is a notable growth increase in the feet of yearlings in the spring and even those with good conformation can exhibit unevenness in foot growth. Differing growth rates are usually the cause although it can sometimes be due to uneven wear. The horse's environment will influence the development of the hoof and should be taken into consideration. For example, the feet of a horse that runs out on wet, marshy land will grow more rapidly and need more hoof trimmed than the horse that is grazed on hilly or hard ground. These horses will need the hoof to protect them against the harder ground and will therefore not need so much trimmed at any one time. Many conditions which occur in the foal are hereditary or congenital and can be the result of nutritional deficiencies in the mare. The nutritive value of the soil is another hidden factor which affects the wear of the hoof. Many faults begin at an early age and are a direct result of poor conformation which has manifested because of mis-management. Much can be done by the farrier working with the owner to correct faults and weaknesses with proper trimming and shoeing before it is too late. A good farrier will begin work to improve poor conformation of the hoof when the horse is still a foal. It is an interesting fact that thrush appears more frequently in foalhood than at any other age.

Throughout the horse's development whilst he is unshod it is particularly important to keep the foot level. Once the young horse begins work he will need a pair of front shoes. Hind shoes at this stage are not entirely necessary, especially if the training includes long-reining, with its risk of kicking. As the horse is ridden and the weight is put on the front feet any weaknesses, such as brushing and over-reaching, will begin to show and the farrier will recognise the need for alterations in dressing the foot and shoeing.

Before calling the farrier the owner must establish what type of attention the horse requires. Whether the feet need trimming, the shoes removed completely or re-shoeing

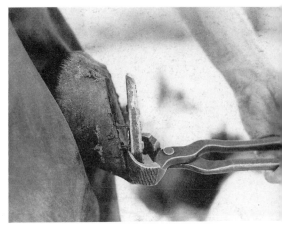

Removing a shoe.

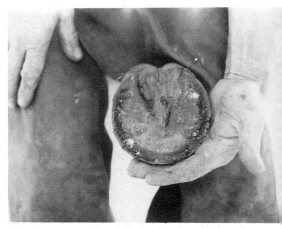

The shoe is removed and the hoof ready to be dressed.

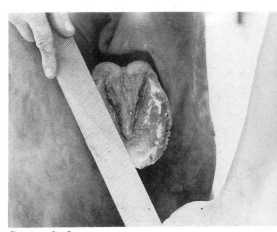

Rasping the foot.

The shoe is nailed on again.

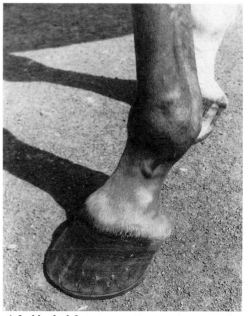

A freshly shod foot.

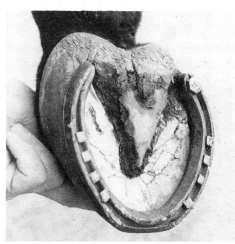

A freshly shod foot with road stud fitted.

with the existing shoes or new shoes. It is more common today for the farrier to travel to the stables having made up the shoes at his forge. If the horse is new to the farrier he may ask you to measure the foot and telephone him with the details so that he may bring the right shoes with him. The foot should be measured across the widest part and an extra inch added. If the length, i.e. from the toe to the heel, is longer, this measurement should be used and an inch added. One of the priorities as far as the farrier is concerned is the need for a clean surface on which the horse can stand. He will also expect the horse's feet to be clean and dry. Any mud should be washed or brushed off and the feet picked out before the farrier arrives so that he can dress the feet as well as possible. He cannot be expected to do a good job if the conditions are unsuitable. As a rule he will carry with him a mobile forge (a gas fire) on which to heat the shoes and adjust them before finally fitting them hot. This process is known as hot shoeing. In cold shoeing the farrier will fit the shoes cold and without making adjustments on the forge. This method is less favourable because it does not ensure as close and tight a fit no matter how well the shoes may have been made, because the farrier has to rely more on the shoe fitting where it touches.

The farrier will need someone to hold the horse while he works. Older horses who have been correctly educated, usually stand quietly whilst being tied up, but young or nervous animals will need someone at their head to keep them as quiet as possible and to teach them the appropriate manners. The farrier will then be able to carry out his job more efficiently, causing the minimum of stress to the horse. As with all the horse's education firm but sympathetic handling is essential, bearing in mind that the young horse will be less patient.

Much groundwork can be done with youngsters in preparation for the farrier's first visit, such as picking out the feet at least once a day. A foal's feet should be handled regularly and it will need two people, one holding him by a foal slip whilst the other teaches him to accept having his legs and feet handled quietly, encouraging him to develop confidence and trust. Before the time comes for the horse to be shod for the first time it is worth preparing him for the banging and tapping on the hoof by using the blunt side of the hoof-pick. The farrier will appreciate all the foundation work that the owner or groom has done and will be able to carry out his work without worrying the horse unduly.

Usually it is sufficient to fit a headcollar whilst holding the horse for the farrier but in some cases, say with a stallion, it may be necessary to fit a bridle. To familiarise himself with each individual horse the farrier may wish to see the animal in action, i.e. led at the walk on level ground, to study the way in which the horse moves, his conformation and general characteristics. His trained eye will immediately recognise the need for any adjustments that might correct and improve the feet. The effect of these alterations often takes some time to become apparent because the hoof is a most critical part of the anatomy and structural changes have to happen gradually. The horse's soundness is the first concern and every care must be taken to see that he goes level. Bad conformation is the chief cause of many faults concerned with the horse's action, such as over-reaching, brushing, dragging the toes, forging or clicking. Faulty hocks often result in the horse dragging his toes. Undernourished and weak horses will be prone to faulty action as will the young, unfit horse who may, for example, brush his hind fetlock joints until the muscles in his quarters are developed and strengthened. This will result in the action of his hind legs being straighter providing, of course, that he does not have any prominent conformational faults such as cow or sickle hocks. The weak horse will be more likely to strike into himself because of his slovenly way of going. Once brushing has reached the stage

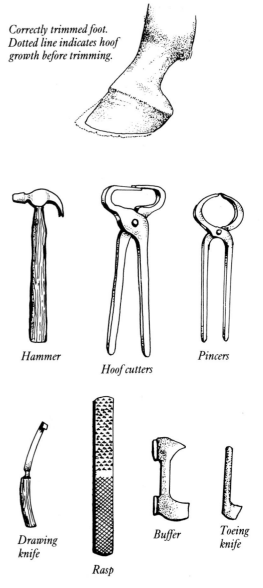

Correctly trimmed foot. Dotted line indicates hoof growth before trimming.

Hammer

Hoof cutters

Pincers

Drawing knife

Rasp

Buffer

Toeing knife

where callouses have developed into hard scabs on the insides of fetlock joints and formed actual protusions it is usually too late to treat effectively. Hind shoes should be kept well under at the toe and the inside branch so that the horse doesn't catch himself with the opposite shoe.

One common habit of the handler which affects the horse's movement is a tendency to always lead the horse from the near side with the horse's head inclined to the left. This is only natural because the horse is less controllable with his head turned away from you. Horses should be taught to lead from either side. Long-reining encourages them to go forward in a straight line. It has been mentioned in an earlier chapter that the horse should be led from the shoulder but it is worth emphasising the importance of keeping the horse's neck and head straight whilst he is being led. With his head inclined towards the left his off fore naturally swings outwards, a fault which will show in the wear of the foot and shoe.

Corrective farriery for common conditions of the lower limb

One of the common ailments of the front foot is a corn. This is usually occasioned by the owner leaving the shoes on for too long or by the farrier fitting the shoe short of the heel. The corrective shoeing for a corn is to take the pressure off the sole, between the bar and the wall of the hoof at the seat of the corn, by fitting the shoe to cover the wall and part of the bar.

Laminitis is a condition which should be diagnosed at an early stage and a close liaison between vet and farrier is important. The farrier's objective will be to arrest any rotation of the pedal bone whilst the toxic blood which has congested in the sensitive laminae causing inflammation must be released. The excess of blood in the coronet will produce an unusually fast growth rate in the hoof, which will be seen in the toe. It will be up to the farrier to keep trimming the toe back and lower the heel thereby transferring the weight from the toe. In chronic cases the pedal bone will rotate until its tip actually pushes through the sole of the foot. A traditional method was to rasp the hoof wall through to the laminae to release the poisonous fluids and to leave as much foot as possible to protect the pedal bone. The theory was that as the horse does not walk on his hoof wall it would therefore not suffer any ill consequences if rasped down, and that the horse's

weight and concussion on the ground would require some compensation. Poulticing may be carried out regularly in conjunction with this treatment in an effort to draw out the infection from the sensitive laminae. Following this line of thought the farrier would not shoe the horse until he was beginning to recover but rather re-educate the hoof back to its normal position by regular trimming of the toe and heel without reducing the length of the foot. The logistics behind this theory are mechanically simple and easy to understand although modern practices favour the use of antibiotics together with a choice of surgical shoes. It must be stressed that each case has to be treated individually depending on the cause, stage of disease, the animal's conformation and so on. Also whether the animal has suffered previously and how it was treated.

Sidebone usually occurs on the outside heel as a result mainly of overwork and is therefore more likely to affect mature horses. The foot will curl under on the offending heel so the shoe should be made thin on the inside, say ½ in (12 mm), and wider on the outside, up to 1½ ins (38 mm), to encourage the foot to grow outwards again.

Ringbone, both low and high, can be treated by the farrier fitting a rocker bar shoe to relieve concussion. Low ringbone is generally more troublesome.

Navicular disease will require the farrier to raise the heel and thin the toes whilst fitting a rolled toe shoe.

Splint – here there is little for the farrier to do besides blanking the inside of the shoe to prevent the horse from knocking the splint with the opposite foot.

Curb – again the toe should be rolled a little as horses with curbs tend to be sickle-hocked and bring their hocks well underneath them. The toe of the hind shoes should be kept well underneath the foot and the heel made thick. No calkin or stud should be used.

Spavins cause the horse to drag his toes, which will need protection.

Stifle lamenesses can be helped by fitting a raised heel, which will also be of benefit to hock problems.

What to look for in the freshly shod horse

A foot which has been freshly shod should obviously be sound and if there is any question at all the animal should be trotted up while the farrier is present. Problems can arise immediately afterwards if the shoe or a nail is ill-fitting and consequently pinching the horse. The angle of the foot should be 45° in front and 50° behind. The heel of the shoe must be long enough to cover the heel of the foot and not be short, otherwise this can lead to several problems. The golden rule of shoeing is that the shoe must be made to fit the foot and not the reverse, which is sometimes the case with cold shoeing. The frog must be kept in contact with the ground to act as a cushion. The shoe should fit in line with the wall of the hoof without rasping too much after the nails have been clenched. The foot must not give the appearance of being 'dumped'. Toe clips should not be hammered too deep into the hoof. The recognised level for the nails is evenly placed about 1½ ins (38 mm) up the hoof. Ordinarily there are seven nails per shoe, i.e. four on the outside where the foot is stronger, and three on the inside. Overall the foot should be well dressed and tidily presented.

If the nail and nail hole are correctly matched clenches should never rise. They should remain in place even when the shoe is worn. The only cause of risen clenches is from excessive movement of the nail in the nail hole. Although the shoe may fit tight the clench will still rise if there is any room for movement of the nail in the hole. The nail size will depend on the size of the shoe and the nature of the horse's work. Sizes range from number four to number seven, the most commonly used being number five. The length of the nail is from 1½ ins (38 mm), used on racing plates, to 3½ ins (90 mm) used on cart-horse shoes. The problem with ready-made shoes is that the nails are all in the same place and cannot be adjusted to suit each individual fitting. It is important that the toe of the hind shoe fits slightly under and as close as possible to the hoof otherwise the slightest space will enable the horse to bend the shoe. Shoes can be easily pulled off when a horse is reversed out of a trailer and the shoe edge catches on the slats.

Common types of shoe.

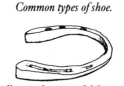

Thinned roller toe shoe – useful for windgalls and navicular.

Plain flat iron with square heels

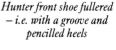

Hunter front shoe fullered – i.e. with a groove and pencilled heels

Hunter hind shoe with calkins on heels – fullered, concave iron.

Racing plate, usually made of aluminium.

Feather-edged shoe for brushing.

Speedycutting shoe, for horses who strike into themselves with their hind legs.

3/4 Shoe, for corns and brushing.

Grass tip, for horses at grass; to prevent cracked hooves.

The owner's farriery kit

Every horse owner should carry an emergency farriery kit to remove loose shoes and rasp untidy edges. The basic kit should comprise a hammer, buffer, pincers, rasp and drawing knife. The owner should be able to remove a loose or worn shoe if he is unable to get a farrier immediately. The farrier will show you how to do this properly, taking care not to tear the wall of the hoof and to remove all the nails. It is advisable to leave the bottom of the foot unrasped for the farrier to trim but do rasp any broken edges to prevent further damage. The farriers' registration law requires that horses may only be shod by a registered farrier.

Hoof care

The need for regular and thorough care of the hoof cannot be underestimated because neglect can be the cause of so many problems which can extend beyond the hoof. It behoves every horse owner to ensure that daily attention to the feet is a matter of routine whether or not the horse is stabled and shod. If the horse is at grass he should have the feet picked out and examined daily for cracks and grass rings. The latter should be regularly rasped off and not allowed to develop. Sandcracks, which result from a change in the weather from hot to cold, can in their early stage be burnt with a hot iron by the farrier to prevent any further cracking down the wall of the hoof. If, however, the crack is established into the coronet, known as false quarter, this will never heal.

The stabled horse should have his feet picked out two or three times a day, say before and after exercise and during grooming. A hoof brush should be used for brushing the feet out, especially if shavings or peat are used for bedding as these materials do tend to pack into the hoof. This will cause the sole to become soft and the frog to rot, encouraging diseases such as thrush.

The use of hoof oils is purely cosmetic and must be regarded as such because the only liquid beneficial to the hoof is water. Water will promote growth of the hoof and to ensure that there is the necessary moisture which is natural to the hoof it is advisable to wash the feet of a stabled horse daily, preferably on return from exercise or after he has been turned out. If the foot is allowed to become too dry it will make it susceptible to cracks, contraction of the foot and stunted growth. In extreme cases where the hoof needs treatment to promote growth the farrier may rasp around the wall about an inch below the coronet the width of his rasp. Having clipped the hair in this area a mild blister may be applied. This is recommended in preference to grooving the hoof in parts. The bars of the foot can be cut out to encourage the frog to touch the ground. Soaking the foot in hot water below the coronet will soften the hoof and encourage it to expand.

Studs and their uses

Horses who are subjected to a large amount of road work are often fitted with road studs, either the type with threads which screw in, or the plug-in sort which the farrier taps into the shoe before it is fitted to the horse. There is a variety of studs for use in competition depending on the job they have to do and in what type of going. Mordax studs are most common as everyday studs.

Competition studs are available for all types of ground conditions. These are mostly fitted in the hind shoes only but in some cases it may be necessary to fit front studs. The horse should not be ridden on a tarmac surface wearing large studs because they will create an uneven bearing surface and damage the hoof and limbs.

Stud holes are made by the farrier when he makes the shoe and can be either threaded or driven. A plain hole is only stuitable for studs which are to be hammered in and left for the life of the shoe. This applies to the hind shoes when road studs are needed. If the studs have to be removed and changed for competition use, the hole will need to be threaded. The holes must be clean and free from grit in order to be able to fit the studs easily. Many people choose to fit small road studs otherwise known as 'sleepers' which can be left in all the time and changed for competition studs as necessary. If the hole is to be left without studs in between use plug it with oiled cotton wool, tight enough to prevent any grit getting into the hole and damaging the threads. With wear it will become necessary to tap out the hole using a tap the size of the hole. The stud kit will need to include a spanner, a tap, a horseshoe nail to clean out the hole and cotton wool, as well as a choice of studs for different types of ground conditions plus replacements.

Routine checks

Every owner should check his horse before it leaves the stable to go on exercise and again on his return to ensure that all shoes are safe and secure and that no clenches have risen. If this is overlooked it can lead to a horse cutting himself on a risen clench or a shoe that has spread, or to losing a shoe and damaging the foot, which can result in the horse being out of work for a while. Even a freshly shod horse can spread a shoe by getting cast or rolling. Risen clenches, no matter how small, can cause aggravating wounds and should be dealt with promptly by tapping them down with a hammer. They are easily detected by running your hand around the hoof. Whenever in doubt about a loose or worn shoe call the farrier because a lost shoe can damage the foot making it difficult to keep a new shoe on. Horses in hard work will need shoeing approximately every three to four weeks and those in light work will need shoes removed and the feet rasped about every four to five weeks. It is false economy to replace nails without re-shoeing as they will protrude above the shoe and wear down very quickly.

CHAPTER THIRTEEN

FEEDING

The principles of feeding

The principles of feeding are listed below for easy reference. They are based on the horse's natural feeding habits and how these can be catered for when he is being conditioned for work.

1. Feed only the best quality feedstuffs available.
2. Feed at regular times.
3. Feed according to the animal's type, size, height, age, health, temperament, condition and state of fitness.
4. Feed at least one and a half hours before exercise.
5. Only use clean mangers and feed receptacles.
6. Maintain a feed list which is kept up to date for each horse's diet.
7. Feed little and often.
8. Always ensure that the horse has access to water before feeding.
9. Processed feedstuffs often have a shorter shelf life once the bag has been opened; a note should be taken of the manufacturer's 'use by' date.
10. Feedstuffs should be examined for quality by smell, texture and colour.
11. Feed something succulent each day.
12. Feed should be offered *damp* but not wet.
13. Any alterations in a diet should be made gradually because the horse's metabolism cannot cope with a sudden change.
14. Bulk food is essential for digestion.
15. Avoid over- or under-feeding.
16. When dampening a feed be sure that it is not left to stand before feeding because it may go sour.
17. The animal should be completely recovered from exercise before being fed.
18. Withdraw feeds which have a protein content when the animal is out of work and feed a laxative diet.
19. Each horse's diet must be individually selected because no two horses' requirements are the same.
20. The horse's teeth must be kept rasped and in healthy order for him to masticate and therefore digest his food properly.
21. Feeding is an art based on a horseman's observation of each animal and being able to recognise the need for a change in the diet.
22. Care should be taken to ensure that horses who are fed in a group get their required ration.
23. Feeding practices should only be disrupted by a situation such as work, competition, hunting etc. and not by human neglect.
24. A feed should be thoroughly mixed before being offered to the horse.

Objectives and nutrient requirements

1. To provide nutrition and make up for the loss of substances in the performance of bodily functions.
2. To supply fuel as energy for the horse's work.
3. To promote the horse's development and growth.
4. To maintain the body in a healthy and thriving state.
5. To repair tissue.

6. For the purposes of reproduction and milk production.

To satisfy these requirements the horse's diet will comprise the following:

Food groups

1. Proteins – vital for maintenance and growth.
2. Carbohydrates (starch, oil or fat and sugar) – energy-producing substances which cannot substitute protein or fulfil their functions.
3. Fibre – woody substance of little feed value. Fibre produces bulk which is essential to the horse. It also stimulates the digestive processes and aids in the assimilation of digestible matter.
4. Minerals – salt, magnesium, phosphorus, potassium, chlorine, sodium and sulphur, plus trace elements required in small doses: molybdenum, chromium, selenium, iodine, copper, cobalt and zinc.
5. Vitamins – vital to good health. Present in natural foods. Horse's requirements comprise vitamins A, B1, B2, B12, C, D, E.
6. Fats – these are present in all foods but in varying quantities. Excess fat causes a build-up of body fat. The 'fatty ratio', recommended as one part fat to two parts protein, should always be quantified when deciding on a horse's diet.
7. Water – again this is present in all foods from as little as 10–12% in oats to 99% in soaked sugar-beet.

Food types

Oats

These are the best energy-producing food, although they are deficient in calcium. They should be clean, hard and plump, heavy and sweet-smelling. Can be fed rolled to varying degrees, i.e. crushed, pinched, crimped or bruised, to improve their digestibility. Ideally they can be purchased whole and crushed as required because once so treated they should

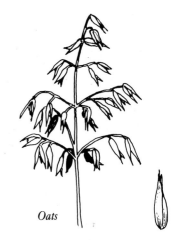

Oats

be used within one to two weeks. Alternatively they may be fed boiled to horses who are not in fast work.

Barley

Barley can be fed rolled in the same way as oats or boiled and used to build up the horse's bodyweight. Care should be taken to ensure that it is cooked with adequate water because the grains will swell during cooking. It must therefore be thoroughly cooked before being fed, at blood heat (not straight from the cooker). The juices of boiled barley are also nutritious to the horse. Boiled barley is not recommended for horses in fast work.

Barley

Maize

Most commonly maize is fed flaked and has a high energy value. It is a fattening food but low in protein and minerals. It can cause overheating if fed in too large quantities and should only be fed to horses who are in hard work. A recommended guide for feeding maize is give no more than 20% of the horse's daily ration.

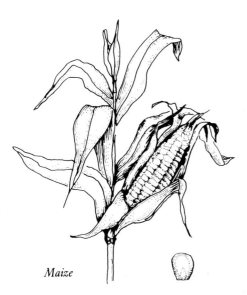

Maize

Bran

As bran is seriously deficient in calcium its feed value is confined to its laxative properties together with being highly digestible although not always palatable. It is becoming increasingly difficult today to find a broad flaked bran. Often fed slightly damped as part of the daily ration to offer bulk and digestibility it is otherwise the main ingredient of a mash. A bran mash, to which boiled linseed may be added for horses who are ill, constipated or at rest, must be fed at a temperature which allows you to mix it by hand otherwise the horse's palate may be scalded and he will be put off his feed.

Nuts, cubes or pellets

The advantages of feeding these pre-packed feedstuffs are chiefly that they offer a standard, balanced diet with all the necessary ingredients for each horse's requirements, are easy to transport and store, do not require mixing and therefore make feeding easy. For this reason they are particularly useful to the many owners who do not traditionally or by virtue of their situation have the knowledge of animal husbandry which the farmer or country person has and are therefore 'playing it safe' by feeding nuts and following the manufacturer's recommendations. These feedstuffs are, however, dry and boring for the horse if he is offered no alternative and he can choke on them if not properly masticated. With so many choices on the market it is important to select the appropriate one to suit each horse's individual requirement and not be tempted to use the same one on all horses as a matter of convenience. They are relatively expensive and have a short shelf life. Once a bag is opened, nuts must be used up quickly as they will absorb moisture from the atmosphere. Stale or damp nuts soon become indigestible and their feeding could lead to colic.

Peas and beans

Both of these pulses are rich in protein, very heating and fattening. They are usually only fed to horses in very fast work, i.e. racehorses, or to animals who are wintered out at grass when they may be fed up to 6 lb (2.7 kg) per day. They should be split, kibbled or crushed before feeding and if they are mixed with other grains, say oats, a ratio of one part beans/peas to two parts oats is recommended.

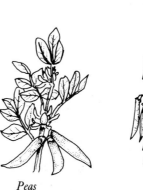

Peas

Field beans

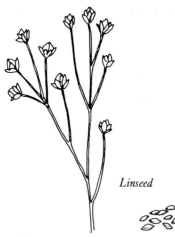

Linseed

Linseed

Linseed is highly nutritious and rich in proteins and oils. It has to be cooked before it can be fed either as a jelly or a tea. Its properties serve to improve the horse's condition and produce a gloss to the coat. A daily quota of 1 lb (450 g) of seed (pre-cooked weight) per average-sized horse is advised. To prepare linseed for cooking first soak overnight in cold water. Add more water if necessary before bringing to the boil and simmering for 2–3 hours. Remember, linseed is poisonous if it is not boiled thoroughly and must be used the same day as it is cooked. Left-over linseed should be discarded on the midden. Linseed tea is cooked in the same way but with more water and is then used in a gruel or mash. Linseed oil is an alternative to the seed and has the same value when fed with roughage and grain. Great care should be taken when cooking linseed because it produces the most difficult mess to clean up once it has boiled over. As a precaution against this a little whole barley or oats can be cooked with it. Ensure the linseed is cooled to blood heat before feeding otherwise it will scald the horse's mouth and put him off his feed.

Molasses

Molasses is a by-product of sugar and as such has a high nutritional value. It is a crude form of black treacle and acts as a useful appetiser, especially for fussy eaters. Liquid molasses is less expensive than molassine meal. As a guide

three to four tablespoons of liquid or one handful of meal per day will enhance the palatability of the feed.

Dried sugar-beet pulp

This is a factory-processed root crop which is dried and produced in either cube or loose form. As such it is dangerously indigestible for the horse and consequently *must* be soaked in at least twice its own volume of water for 24 hours before it is offered to horses. It must be fed fresh once it is soaked and not be allowed to stand for more than 48 hours in cool conditions (less in warm weather) because fermentation can occur. It is a useful source of energy and roughage which is both highly digestible and one of the best feeds for increasing body weight. It is also very laxative when used in excess.

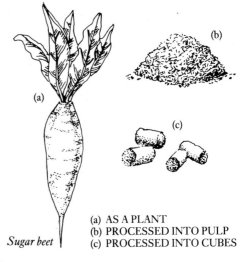

Sugar beet

(a) AS A PLANT
(b) PROCESSED INTO PULP
(c) PROCESSED INTO CUBES

Carrots

Apart from being very palatable food for horses, carrots are relatively high in vitamin A, B and C. Before feeding they should be washed thoroughly to ensure they are free from slugs and insects and not rotting, and sliced lengthways to avoid the risk of choking. Take care to store them in a dry place because once damp they will soon form mould and rot. About 1–2 lb (450–900 g) a day is enough to make a feed succulent and encourage the horse to masticate his food properly, thereby improving digestion.

Apples

Again these offer the horse a succulent and appetising addition to a feed and will tempt the delicate feeder. Chop into quarters before feeding. Never feed any which are starting to rot.

Chaff or chop

Chaff is chopped or shredded hay; Chop usually refers to chopped oat straw. Sometimes a mixture of chopped oat straw and hay is fed, which adds bulk to a feed. It is produced by a machine called a chaff cutter. Use good hay and mix well into the rest of the feed. Chaff and chop make the horse masticate his food and thereby aid digestion. The quantity fed will depend on the nature of the horse's work, e.g. if on fast work his requirement may be as little as a double handful and no chop; on the other hand horses in slow work or at rest can have as much as a scoopful in each feed.

Eggs

Raw eggs are often fed to horses in top condition and fast work as an extra conditioner. Feed only for short periods, i.e. four to six per day for three to four days.

Milk

Powdered milk is often fed to foals and weanlings. Milk can also be offered to horses in training, either in powdered form or naturally. In the normal diet up to 1 gallon (4.5 litres) a day can be fed.

Beer or Guinness

This is traditionally used for horses in fast work, such as racehorses, and has proved to be a useful tonic. The recommended quantity is about two bottles a day for three to four days before a competition.

Cod liver oil

This is a natural and excellent tonic which also promotes growth of horn, tissue and bone. One tablespoon twice a week, mixed thoroughly into the feed, will generally be sufficient for the healthy adult horse. Do not be tempted to feed more without veterinary advice because this will often imbalance the horse's diet which is often the case with many overfed supplements and additives. It may be necessary to introduce the delicate feeder to it gradually by disguising it with carrots, apples or molasses. The alternative to liquid cod liver oil is Codolettes (small spiced nuts), which horses sometimes prefer.

The following list is given as a guide to the percentage of each ingredient which can be included in a ration for horses.

	% air dry feed
Oats	90
Barley	15 (fed cooked)
Wheat	0
Rye	0
Maize	30 (flaked or cracked)
Field beans	10 (boiled or cracked)
Linseed	10 (fed cooked)
Wheat bran	10
Dried molassed sugar beet pulp	15 (fed soaked)
Molasses	10
Soya bean meal	15
Carrots	75
Fish meal	10
Dried skimmed milk	15
Dried yeast	0.2
Grass meal	50 (fed damp)

% PROTEIN   RATIO

MIX: 10.0 $8.5 - 6.0 = 2.5$

8.5

HAY: 6.0 $10.0 - 8.5 = 1.5$

TOTAL $2.5 + 1.5 = 4.0$

FOOD REQUIREMENT = 2.5% OF LIVEWEIGHT = "X"

AMOUNT OF MIX = "X" $\times \dfrac{2.5}{4}$

Hay

The most economical way of buying hay is directly off the field in June or July as soon as it is baled. The reason for the saving is that contractors' charges are avoided if the hay does not have to be carted and stored elsewhere. Buying this way cuts down on labour and transport but requires that storage space is made available at a convenient site near the stables. The price will increase the longer hay is stored with a contractor, becoming most expensive after Christmas and in the spring. If you have your own transport and labour, you will also have the benefit of choosing the hay yourself. Once stored, hay will lose weight as it dries. The top and bottom layer of hay is generally unfit for horses: the top becomes dusty and the bottom layer damp. A layer of straw under the bottom layer will prevent this.

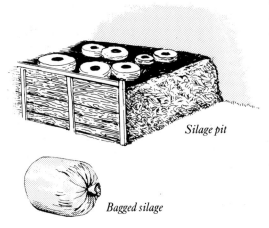

Silage pit

Bagged silage

Once the local merchant or, better still, farmer knows what you need for your horses he can be asked each year to choose the best hay for you and advise you on collection or delivery. He will usually leave out any dusty or mouldy hay, which could be fed to cattle or sheep. Hay should not be fed to horses in the year of its harvest so you will have to order enough to last from one year to the next before starting on the new hay. The change from old hay to new should be made gradually, over a few days, mixing some of each so as not to upset the horse's digestion. New hay fed

Bagged Haylage

within four to six months of harvest can cause digestive disorders and inflame the legs. Unless space is limited it is uneconomical to buy quantities of less than a ton at a time.

Identifying good and poor quality hay

Hay should always look and smell fresh with no signs of damp or mould. There should be a minimum of dust; some dust is inevitable but an excess is harmful to the horse's respiratory system. A bale of hay should fall apart easily and be greenish to light brown in colour. Stabled horses will be more selective when eating hay and leave any inferior hay. Hay that has become overheated in the barn will often develop a dark brown colour and a sweet tobacco odour. Although horses like this, much of its feed value has been lost. Recognising good and bad hay takes years of experience so it will pay the novice to seek expert advice before he buys or makes hay of his own. There is no point throwing good money after bad. Hay made late in the season will have lost much of its feed value as most of the seeds will have fallen out. Hay that is light in colour may have been rained upon or left out too long during its making. Hay that is two years old or more will also have lost its feeding value.

Types

1. Meadow hay – made from permanent pasture.
2. Seed hay – made from temporary pasture, i.e. one-, two- and three-year leys, as part of a rotated crop.
3. Lucerne (or Alfalfa as it is known abroad) – is high in protein and high yielding, i.e. five cuts per year.

Meadow hay mixture includes cocksfoot, timothy, sweet fescue, sweet vernals, crested dogtails, red clover, rye grasses. Yorkshire fog and smooth meadow are inferior grasses. Common meadow barley and common rushes are undesirable.

Seed hay mixture includes rye grass, cocksfoot and clover.

The nutritional value of any hay relies on variables such as the types of grass present and the weather, and a combination of these two results in an inconsistent product. Traditionally horse hay is made in June or early July when it is at its best. Depending on the weather, it will be mown, dried, turned once, twice or even three times before being rowed up for the baler. Once baled it can then be stacked and carted provided it is well aired and not in danger of overheating. Drying hay in England is a risky business because of the temperate climate; sometimes it will need drying artificially. When examining hay look for a good colour without extremes, a sweet smell and crisp texture. A bale should separate easily when opened and not be heavy. A brownish-yellow colour indicates that the hay has been mow-burnt and is therefore unsuitable. If the hay is brown to black it will be mouldy and rotten. A yellow to bluish green colour is just acceptable. The ideal colour is a bluish green because it suggests that the hay has retained the carotene content. Unrecognisable seed heads suggest that the hay has been on the ground for too long. As a rule, horses in fast work will require seed hay whilst meadow or soft hay is more suited to general-purpose horses and ponies and those out of work. Whatever the animal's requirements it is false economy to feed anything less than good quality hay because inferior hay will soon affect the horse's respiration, often with irreversible effects.

A Dutch barn or large high building is most suitable for storing hay but a stable loft needs to be spacious to avoid the hay becoming overheated. Care should be taken when storing to allow plenty of room for the hay to air and that there is ample ventilation. It is preferable to store hay for twelve months before use, but that will depend on its quality and state when stacked. An approximate good weight for a bale of hay is 30–40 lb (13–18 kg).

Hay can be offered in a variety of ways and much will depend on personal circumstances and preferences. However, no one can afford to waste hay on any horse so it is prudent to consider how effective the manger or hay rack is in terms of spillage and wastage. Racks which are secured to the wall at a height of approximately 6 ft (1.8 m) will mean that the horse has to reach up to pull out the hay. The same applies to haynets to a lesser extent but both methods have the sometimes undesirable effect of developing the underside muscles of the neck, the opposite to what most equestrians are trying to achieve. Feeding hay on the floor can be extremely wasteful unless very small portions are fed at any one time. This would involve a lot more work but it is the most natural way for a horse to eat. The ideal method must be a low, deep manger which enables the horse to eat in a natural position without wasting too much. Whichever method is used the hay should be shaken up well in the hay barn to examine it and if necessary to damp it. Many people today soak the horse's hay as a matter of practice because hay has been found to be the cause of so many respiratory disorders; the most common of these is referred to as COPD (Chronic Obstructive Pulmonary Disease).

Watering

Water is a variable commodity depending on its source – for example, mains water in invariably treated. Soft natural rain water is most desirable, but not always practical to catch in a tank and offer to horses, especially in a large yard. Horses who are used to soft water, such as the Irish, take time to get used to the hard mains water in England. The effect will be seen in their condition, which may temporarily deteriorate, and owners should be aware of the cause. The state of the skin and coat will be affected. Likewise, the benefit of giving soft rain water to horses will be apparent in the glossy condition of the coat after a time.

Particular care should be taken in watering horses, especially those in hard work. They should not be allowed sudden access to a large amount of water immediately before or after exercise. Instead a constant supply is favoured so that the horse is not allowed to become dehydrated. A few people still water horses at intervals throughout the day in preference to leaving water in the box, but this is less practical especially in a large establishment. Automatic water bowls or water buckets filled several times a day will ensure a constant fresh supply. Buckets must be completely changed for fresh water at least four times a day and some horses may need two buckets overnight to ensure that they do not go without. Clearly if water is not constantly available a horse will be more likely to drink large quantitites in one go, to make up. This can be the cause of colic, especially if taken just before or after feeding and exercise. Excess water immediately after feeding will wash the food through the stomach before it is properly digested.

Rules of watering

1. If water is not constantly available it should be offered at least six times a day. This may apply to an injured horse who has to be tied up for long periods, or to animals in stalls.
2. Water vessels must be kept scrupulously clean as water will absorb atmospheric odours making it sour to drink. The vessel will stain easily and taint the water.
3. Avoid horses gaining access to old troughs and ponds where the water will invariably be stagnant.
4. Water should always be made available before feeding.
5. Whenever possible it is advisable to take one's water supply to competitions to avoid a change and risk infection from an unknown source.
6. During long periods of exercise, such as hunting and long-distance riding, allow the horse a drink to wash out his mouth but do not let him drink a large amount.
7. When a horse returns to the stable after a long time away without water, such as on exercise or travelling, he should be offered water to which has been added a little warm water to take off the chill. He should then be offered a small amount at frequent intervals until his thirst has returned to normal. Never allow him large quantities of cold water at this time.
8. Water that has become fouled should be changed immediately.
9. Ensure that each horse is watered according to his individual requirements. This includes allowing for those with a large thirst.

By nature a horse drinks and eats little and often to suit his digestive system, so it is dangerous to allow him an excess of food and water simultaneously. He will be more comfortable with an adequate supply of hay and water which he can consume over a period of time. In his natural state he is able to graze and drink simultaneously without ill effect but this does not follow when his diet is changed to concentrates.

Water bowls

Automatic water bowls allow the horse a constant supply of fresh water 24 hours a day, providing they are functioning normally without any blockages such as are caused by bedding and fodder. They must be checked regularly for breakdowns and blockages otherwise the horse may have been without water for a unknown period before the fault is detected. This could prove injurious to the horse. The disadvantages of water bowls are that it is difficult to administer medicines in a small amount of water as opposed to a large bucketful. It is also impossible to measure and control the water intake of each horse, which could be too much or too little.

Outside watering

If a natural source of running water is available it must be easily accessible. If there are only steep banks leading to a stream, an access should be dug away to allow the horse to approach safely. If there is sand at the bottom of a stream horses must be checked regularly for sand colic or scouring which can be caused by the horses taking in sand when they are drinking. Scouring, in the extreme, can prove fatal so owners must be extra vigilant where there is a running stream.

A water trough with a piped supply is ideal and can be kept clean and fresh. It is important to check all water supplies regularly but particularly with artificial supplies. These should be looked at twice a day when horses are out, and especially in winter when they may freeze. Both the ice and pipes will need thawing. Where possible, pipes can be lagged to prevent frost damage. Troughs should be sited so that they are easily accessible, and surrounded by stone or gravel to prevent a mud bath. A concrete apron is dangerous in the winter when frosty.

Difficult feeders

Horses who do not eat up willingly are a headache to any owner. Those in competiton work are a particular worry as they are difficult to maintain in a good condition. Difficult feeders are usually highly strung animals with a nervous disposition. If a horse becomes unsettled when he is eating and does not finish his feed readily there may be a pathological reason for his loss of appetite. It may be prudent to call the vet if there is no obvious reason, such as a change of environment, work, company or a competition, indeed any situation which is stressful to the horse. The next thing to check is the horse's teeth which are often the cause of the horse being unable to masticate and therefore digest food properly. Another cause can be a blood disorder which the vet will diagnose and treat.

Once these factors have been eliminated the owner will have to be patient and consider different ways of encouraging the horse to eat. Shy feeding can be caused by prolonged stress, say from over-training, which has left an animal psychologically disturbed. This will often take a long time to overcome. The horse's confidence must be regained and he must be kept as quiet as possible with the minimum of stress in his routine. The owner will need to establish at which time of day or night the horse eats best and feed the largest quantity of feed at that time. Many horses feed better at night when the yard is quiet. It is often found that a companion, such as a goat, will have a calming influence on an unsettled horse and encourage him to eat.

Mangers and all feed containers should be kept thoroughly clean to prevent a stale odour developing. This will quickly put a horse off his feed. Likewise all left-over food should be removed from the manger before another feed is offered. Always practice the golden rule of feeding little and often during the day. Apples, carrots, molasses/black treacle, and grass mixed in the feed will often encourage the shy feeder. Only quality food should be fed. Inferior feeds give off a bad odour and are harmful to the horse. Horses in a poor condition or ill-health should be fed very small quantities of the best quality food and, if in doubt, veterinary advice should be sought. By their gregarious nature some horses are encouraged to eat when they can hear or see other horses. For this reason it may be worth fitting bars into the walls of adjoining stables between mangers. Often it is helpful in relaxing a nervous horse to turn him out to grass for a while to settle him before bringing him in to feed undisturbed.

The horse's likes and dislikes should be established too. Some prefer being fed dry food to wet, and vice versa. There may be different foods which they like best and a diet can be devised with this in mind. Many horses dislike mashes and gruels. Where there is a lack of appetite other supplements can be offered such as boiled linseed, barley, milk, eggs, ale and sugar cubes. A sod of fresh turf is popular with most stabled horses and can be fed on the floor.

It is important to recognise the difference between a genuinely difficult/shy feeder and those animals who are off their food temporarily due to a stressful situation such as at weaning, during a competition or on being parted from a companion.

Bran mashes

The traditional practice of feeding a bran mash once a week, usually after a competition or a day's hunting, can also be beneficial for horses who are out of work through injury or illness and must be stabled. The regular bran mash comprises 3–5 lb (1.3–2.2 kg) of bran to which is added 1 oz (28 g) of common salt or Epsom salts. Boiling water is then poured over to produce a crumbly texture and the whole stirred thoroughly before covering with a sack and leaving to stand for about half an hour or until cool enough to feed at blood heat. A bran mash should not be too wet and sloppy as this will make it unpalatable. Other variations of bran mash have the same basic ingredients but with the addition of boiled linseed, boiled barley or boiled oats. A few crushed oats can be added for those horses who are reluctant to eat a bran mash by itself. Black treacle or molasses are useful supplements to a bran mash and will often encourage a shy feeder. Some horses prefer a cold, wet bran mash in summer.

Over-feeding

It is very dangerous to over-feed any horse as it can lead to many problems not least when he is stabled. Early symptoms can be digestive disorders resulting in colic, filled legs and joints, particularly the hind joints. The latter symptoms may lead to sores and cracks, sometimes in the heels. A more serious condition which can be caused by over-feeding is laminitis or 'fever of the feet'. This condition relies on early detection and treatment if it is to be stopped before it reaches a chronic stage. Heating feeds, such as barley, beans, flaked maize or oats, are particularly dangerous to over-feed. Whenever a horse is subjected to a change of his exercise routine and confined to his stable through injury etc. he must have his diet adjusted appropriately. Heating feeds such as those mentioned should be reduced at once and a laxative diet fed. All concentrates with a high protein content should be omitted from the feed until the horse resumes his normal work schedule.

Under-feeding

The obvious symptom of a horse being under-fed is an apparent loss of condition. There will be a weight loss on the hindquarters, neck and loin areas as well as a tucked-up effect on the belly. Bearing in mind that each horse must be treated individually the reasons for a weight loss must be established and a new diet devised. First of all the horse must be examined to determine that there are no other symptoms of ill-health. If there are, the vet must be called in to advise on a form of treatment to coincide with a revised diet. As well as a lack of food, over-work, excessive stress and poor quality food can all cause weight loss. Conformation can also be deceptive and a horse with a long, thin back will probably always be difficult to keep in a good condition, as will a horse with a nervous disposition. Never forget that the horse has a comparatively small stomach and needs small feeds at regular intervals if he is to thrive. Serious undernourishment can often take many months to overcome before the horse responds physically to a correct diet.

Change of diet

Any changes to a horse's diet must be made gradually if the horse is to benefit fully by converting foodstuffs efficiently. A change of diet will take several days before it is apparent in the horse's condition. As always the food should be mixed thoroughly. Horses are naturally suspicious about new smells and tastes and may have to be persuaded gradually. Any changes or administrations of supplements, additives, wormers or other medicines are best given in the evening feed. If necessary a smaller feed may be given when introducing something different. Sugar-beet pulp is usually relished and therefore useful when mixing in supplements and additives. A handful of food can be mixed in with medicines before adding to the feed which will ensure it is thoroughly dispersed and disguised as much as possible. Whenever medicines are administered in the feed, nuts are best left out because some substances stick to them, making it easy for the horse to separate them out from the rest of the feed. Before offering a feed with medicine, or a new supplement or additive, be sure that the horse is expecting a feed or he may refuse it from lack of appetite, especially if he is a fussy feeder. A shy feeder may be encouraged to eat new feedstuffs if they are added once he has started his meal; likewise a layer of carrots or apples or molasses sprinkled on top of the feed once it is in the manger will encourage a horse.

Preparing a feed

The ingredients for each horse feed should be put into a large bucket and mixed thoroughly by hand or with a stick to ensure that they cannot be separated easily by the fussy feeder, who may spend time picking out the food he prefers. Some people prefer to mix the food in the manger but if this is done the horse must be made to wait on one side and not push in until the groom is happy that the food is completely stirred. In large establishments such as studs, a barrow is often used for feeding when all horses are having the same meal, perhaps nuts or a bran mash. While this method is time-saving and simple the groom must be sure that the food is thoroughly mixed by hand or with a shovel before it is fed. A feed barrow must never be used for any other purpose.

Cooked food

Foods such as barley and oats can be fed boiled whole to horses who need to gain weight. Cooked food is also more digestible and often more palatable. Linseed grains have to be soaked and boiled before being fed. Oats and barley are less heating once they are boiled and can be fed to a horse on a laxative diet. They can also be added to mashes and make an appetising feed for a tired horse. They will need to be boiled for 4–5 hours until the grain can be pinched between finger and thumb. Special heavy-duty boilers are available for cooking grain but they are expensive

and economical only for larger yards. Boiled food should not be fed to horses who are in fast work as it can damage their respiration.

Buckets and cooking vessels used for boiled feed should be scrubbed clean after each use. Left-over food should be thrown away and mangers wiped clean otherwise stale food will soon become sour and contaminate containers. This will soon put horses off future feeds.

Gruels

Linseed tea or oatmeal gruel are useful feeds for tired or sick horses as they are very nourishing. A gruel or tea is made up of about 1 lb (450 g) of meal to 1 gallon (4.5 litres) of water. It should then be boiled and left to stand until it is cool enough to feed, ideally at blood heat.

Exercise, diet and health

In order to keep the horse in good health his diet and exercise must be suitably designed to maintain him in a thriving condition whether in a state of fitness or rest. No two animals are the same and what may suit one horse will not necessarily suit another. Careful observation should reveal the need for changes in the horse's exercise and diet as his fitness progresses or as his body's maintenance requirements alter in a state of idleness. Without properly controlled exercise for the working horse his fitness will not improve although he may be perfectly healthy. Lactating mares and youngsters must have regular, i.e. daily, exercise to enable their bodies to convert foodstuffs efficiently and to their fullest advantage.

Each horse will have a different requirement in terms of exercise for a specific training purpose. The character of each discipline will dictate the nature of the work required to produce the type of condition necessary for that competition. In order to create a training programme for each horse these requirements together with an understanding of how to interpret them for each horse are needed.

The horse's health will at all times depend on the right amount of exercise for his situation. If a horse's exercise is limited whilst he is corn-fed and stabled his constitution will suffer. Even a horse at rest must be exercised – albeit a short walk in hand. The horse's metabolism will not convert feedstuffs efficiently if exercise is lacking and as a consequence his health will suffer. Early symptoms are filled legs, azoturia and colic. If a horse must be confined to his stable for reasons of injury or ill-health his feed must be adjusted accordingly, i.e. from a diet of concentrated feedstuffs to a laxative diet which will serve to prevent his blood from becoming overheated.

The art of feeding

As with any aspect of horse management experience only comes with years of practice, but, even so, not everyone will develop the skill of a good feeder. The key lies in having an eye for a horse's condition, knowing each horse in his various states of fitness and recognising the need for a change in diet. This must then be put into effect without causing the horse any undue stress without him going back in condi-

tion or his fitness affected. A good feeder will, by careful observation, know at a glance when the horse needs his diet adjusting and this can be for reasons other than physical. The strains and stresses of travel and competitions will have their effect on every horse but some cope better than others. The good feeder will recognise this and be able to improve the situation through feeding without interfering with the horse's state of fitness.

The old-fashioned stud groom was often a good feeder, particularly alert to the needs of the horses in his charge and, as is less common today, his budget was less restrained which afforded him the best quality food and facilities available. It has to be said too that in his day, there was more time for the horses and owners which sadly is less available today.

Anyone can throw a bucket of food into the manger, and a lazy groom may even put fresh food on top of stale which will eventually put off a horse from eating. Feeding is not for generalising but must be given careful consideration if the horse is to be kept in a good condition, thriving on his diet and converting the food efficiently.

GRASSLAND MANAGEMENT

Assessing the land

The effectiveness of the grazing area available can be greatly enhanced by good management, having consideration for the climate, soil type, aspect and elevation. A grass sward is a constantly changing collection of plants and any changes should be well managed to ensure that the sward will maintain a high proportion of the productive species of grass. A badly managed sward will eventually revert to the indigenous vegetation of the area, which will be both unpalatable and unproductive.

First assess the existing situation:

1. Check the soil acidity. The ideal is a pH value of 6.0–6.5. Anything below 5.5 will require the application of lime otherwise there will be an uneconomic response to fertilisers or seeds.

2. Check the phosphate and potash levels. Anything less than index 2 will require supplementation.

These two tests are done by soil analysis, a service which is usually available free from any reputable fertiliser supplier, but an independent assessment is also obtainable from the Ministry of Agriculture at a small cost. Lime application will not be an annual requirement but the levels are certainly worth having checked every five years. Phosphate and potash are plant foods and depending on the level of production will probably need to be replaced annually at the rate of 40 units/acre of phosphate and 60 units/acre of potash, but here again need checking every three years.

3. Check the drainage: wet patches encourage weed grasses and rushes, shallow rooting and poaching. Ensure all ditches are clean and free flowing and that the outfalls from any existing drains are clear of obstructions. Drainage of consolidated land can be improved by 'sub-soiling'. This service is available from agricultural contractors and is a means of loosening up the sub-soil and relieving the compaction.

4. Check the weed problem. Weeds compact with grass and they can be controlled by herbicides. Advice is available from any agrochemical supplier. Buttercups, dandelions, rushes and plantains are indicative of poor fertility and may require an all-over treatment, whereas docks, nettles, thistles and chickweed often grow in patches and can be 'spot-treated'. Do beware of ragwort. It has the appearance of a 'rosette', can cause digestive upsets and is quite toxic for some weeks after spraying. It should not be grazed or cut for hay until the foliage has disappeared. Ragwort, being a biennial plant, sheds seeds which germinate over a long period and therefore it may be necessary to spray it in two successive years. Spraying can be damaging to clovers in the sward and if you are not re-seeding it may be important to choose a spray less damaging to clovers at the expense of effectiveness to the weeds. Hay fields are best sprayed in April or May with a repeat dose in August or September, and grazing areas in May/June or August/September with a repeat dose the following season. Do not graze any area within fourteen days of spraying. Weeds will reappear in the course of time but small areas can easily be spot-treated which may help to keep on top of the problem.

Poison ivy

Privet

Yew

Ragwort

Foxglove

Deadly nightshade

Rape

Assuming these preceding corrections have been made where necessary and the sward is still unproductive, then consider the need for re-seeding. It may well be an old sward of unproductive species and the time for rejuvenation is at hand. If your field is level and uncompacted a new crop can be drilled into the new sward. This is known as 'direct drilling'. However, an unlevel, compacted sward with a thick mat of old grass, molehills and such like will need to be drilled by the 'minimum cultivation' method. Only in very extreme cases will one have to resort to ploughing. It is advisable to contact a reputable seed merchant with regard to your seeds (which will be discussed shortly), for he will put you in touch with an approved contractor with the correct equipment to carry out your re-seeding. After all, the seedsman has an interest in seeing that the seeds he sells you are going to be properly planted.

With the direct drilling method the old sward will be sprayed with Paraquat and left for ten to fourteen days for all the vegetation to die, during which time any fertiliser or lime could be applied and the re-seeding will take place thereafter. Roll or harrow after re-seeding to close the seed slots if necessary. The best time to do this is September or October whilst the soil is still warm yet is not so likely to suffer drought.

The minimum cultivation method may require two sprays at fourteen-day intervals if the 'mat' is very thick. When vegetation is dead the land is rotivated or shallow cultivated to incorporate the trash into the top soil. It should be sown immediately before the soil has time to dry out, and rolled immediately, twice if necessary, as the conservation of moisture is of prime importance for germination.

In both cases, where perennial broad-leaved weeds are a problem, an additional weed-killer may need to be added to the

Paraquat. Old grassland normally harbours slugs so it is advisable to ask your contractor to mix some slug pellets with the seed mixture at sowing.

Seed mixtures

Your seed merchant will advise you on these but it is worth knowing that they can be made up to choice depending on the use required of the grass, e.g. grazing (cattle, sheep and/or horses) or cutting for hay and the soil type, climate, etc.

Commercial varieties commonly available:

Italian and hybrid ryegrass – biennials, leafy, early growing, aggressive but not persistent and therefore mostly used as a 'nurse' crop for other grasses in a permanent situation.

Perennial ryegrass – persistent over several years, early growing, leafy, palatable, lacks mid-summer production and needs abundant moisture.

Cocksfoot – deep-rooted, withstands drier conditions, coarser and less palatable.

Timothy – non-aggressive, difficult to establish, does not get going till its second year, very palatable, good mid-summer production.

Meadow fescue – good companion to Timothy providing bulk of production later in the season.

Tall fescue – not so palatable, taller, but provides early and late grass.

These varieties are normally in short strains for grazing and tall strains for cutting (hay or silage) and there are often medium strains suitable for either purpose. Your seedsman will advise you also on the availability of other grasses and herbs such as smooth-stalked meadow grass, crested dogstail, chicory or yarrow, all of which will enhance the palatability and give a good bottom to your sward. Grasses are normally accompanied by white clovers (about 2 lbs/900 g seed/acre) which as a legume have the benefit of producing nitrogen in the soil to the benefit of the grasses; they are low growing and quickly develop a dense palatable 'mat'.

Perennial rye, good feed value, hardy except in frost.

Cocksfoot, nutritious. Best short for palatibility. Thrives on dry land.

Timothy. Nutritious, popular with horses. In leaf until late June.

Yorkshire fog. Low feed value.

145

*Sanfoin.
Bright green high
starch and protein
content.*

*Meadow fescue.
Very similar to
sheeps fescue.*

*Clover hay.
Brown in colour.
Very rich, very high in
lime and protein.*

*Purple moor.
Grows well on wet
land. Adds density
to a crop.*

*Lucerne/Alfalfa.
High in protein and lime.
Grown for haylage.
Bright green.*

*Sheeps fescue.
Hardy grass; grows well
on acid soil deficient
in phosphates.*

*Meadow hay.
Good mixture of soft
green grasses. High
in starch and
protein.*

146

Do's and don'ts of paddock management

Don't overgraze – grazing too short can damage the growing cells of the grass and can kill out productive varieties allowing weeds to re-appear.

Don't undergraze – grasses will go to seed if not grazed adequately, becoming coarse and unpalatable and taking nutrients from the soil unnecessarily.

Do roll at least once each spring to consolidate topsoil in order to conserve moisture for the summer.

Do harrow occasionally to spread the dung, rip up any dead vegetation and aerate the sward. This will normally be necessary after the spring rolling and before the onset of winter. Avoid harrowing during periods of drought.

Managing grass growth

As a result of our climate, even with the best orientated management and seed mixture, the maximum grass production will occur during May and June. In order to absorb this the stocking rate (number of animals per acre) must be increased accordingly. If this is not feasible then some thought must be given to conserving the surplus grass as hay to avoid undergrazing. If an area is to be cut for hay do not graze it after mid-April and ensure that it is securely fenced off until the hay is cut and carted. The rate and amount of grass growth can be manipulated by the application of nitrogen fertiliser. Nitrogen is a very quickly absorbed plant food and is therefore applied as and when required, i.e. early spring and monthly thereafter at ½–1 cwt (25–50 kg) per acre each time, according to the growth rate required. It must be appreciated that it is no substitute for rainfall and no amount of fertiliser will make the grass grow in a drought. If an area is to be enclosed for hay, then apply 2 cwt (100 kg)/acre to this area in April and then no more until after the hay is taken. In mid-summer some less-palatable grasses, particularly if grazing is light, will tend to be

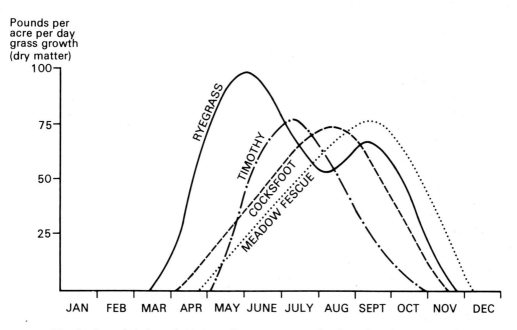

Typical varietal variations of grass growth showing how a good seed mixture will create a full seasons' supply of grass under average climactic conditions

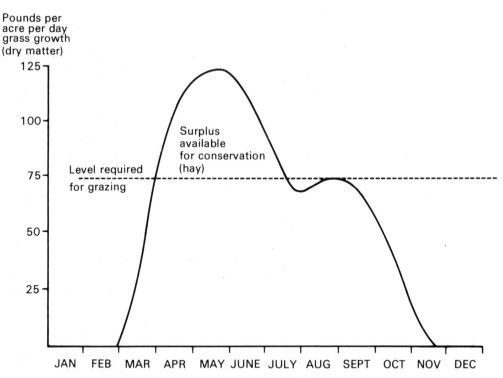

Pounds per acre per day grass growth (dry matter)

125

100

Surplus available for conservation (hay)

Level required for grazing

75

50

25

JAN FEB MAR APR MAY JUNE JULY AUG SEPT OCT NOV DEC

How the use of nitrogen fertiliser in the spring produces growth surplus to grazing needs to produce grass for haymaking.

left uneaten and will start to go to seed. These should be regularly 'topped', i.e. mown off at about 6 inches (15 cm) to maintain their vigour. Don't forget, grass is not just 'something that grows', it is a crop in its own right and must be treated as such. If treated well it will treat you well.

Hay making

When considering the making of hay, the time of cutting is of prime importance. You can either opt for quality at the expense of quantity or vice versa, or, as is more usual, take the middle course. Whichever avenue you choose you are at the mercy of the weather so try and combine your efforts with a predictable spell of fine weather. Grass will be at its peak of quality by about the second or third week in May, depending on preceding weather, seed

mixture, aspect and elevation. The seed heads will not have appeared and the protein and digestibility factors are at their highest. The disadvantage, however, is that the grass is very young and sappy, hence it will take longer to become dry enough to bale and the resulting weight loss will be considerable; thus the yield will be low. At the other extreme, by early July the grass will have gone to seed, be at its maximum bulk and be relatively dry and fibrous. Although taking less time to become dry enough to bale it will make coarser, less nutritious and less palatable hay. June would therefore appear to be the optimum month and many old farming folk consider that the best hay they can make is 'Wimbledon hay', i.e. made during the Wimbledon fortnight. If your pasture is new, modern strains of grass do tend to be slightly earlier maturing than older ones, so aim to get all your hay into the barn at least in time for the semi-finals!

Do make prior arrangements with agricultural contractors well beforehand if you do not have the necessary equipment to hand as timeliness is of paramount importance. Do not mow too short: leave at least 1½ ins (3–4 cm) of stubble as this will help to aid a speedy regrowth of your grazing afterwards and help to prevent the sward from drying off in a drought. If the weather is fine after the mowing, which can be done in the rain, turn or ted the grass at least twice a day. If rains threaten, leave the mown swath untouched and it will not deteriorate in this condition for several days. The tedding and/or turning process will depend on the weather and crop density. A light crop in hot weather can be ready to bale in twenty-four hours but this is the exception. A heavy crop in humid weather may take ten days. Examine the crop frequently in various parts of the field and when it is dry enough it will rustle when shaken and be bristly to the touch. If it is baled in good condition it can be carted and stacked immediately. Hay of dubious dryness will 'sweat' after baling and the bales may become warm after a few hours. Do not put this into a barn or you risk a fire. Leave it in the field until the bales have cooled, say twenty-four hours at least. Don't be afraid to seek advice if you haven't made hay before. Hay is a variable commodity and even experienced haymakers don't always get it right.

Field boundaries

Whether your land has hedges, fences or walls or any combination thereof you have a moral responsibility to make them safe for your stock and a legal responsibility to prevent your stock from straying. Hedges should be trimmed annually to encourage them to remain bushy and not tall and spindly. New hedging plants can be planted in gaps or these can be filled with wooden rails. Hedges that have grown too tall and thin to be stockproof can be cut and laid but this is a job for a professional hedge-layer.

Inspect your hedges frequently and remove any poisonous plants, e.g. deadly nightshade, laurel, yew etc. Your hedge is an important

Dangerous fencing.

feature; not only can it retain your stock, it is invaluable as a windbreak. If your hedge is not 100 per cent stockproof you will require a fence in front of it. Here you have a variety of choice, but, as is usual, the best is the most expensive. This will be a post and rail fence which will look fine when it is new but will require constant care and attention in the course of time and its maintenance too can be expensive. But it is stockproof and, equally important, it is very safe for horses. Hardwood, such as oak, is more expensive than softwood but will last longer and not be so liable to be chewed.

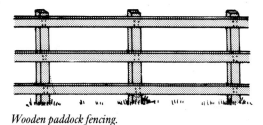

Wooden paddock fencing.

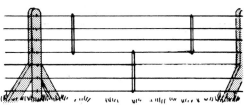

New Zealand fencing.

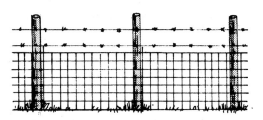

Wire mesh netting with barbed-wire.

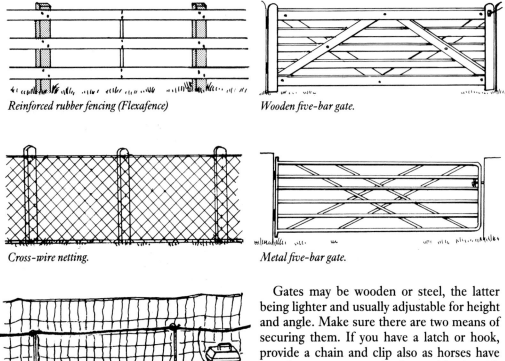

Reinforced rubber fencing (Flexafence)

Wooden five-bar gate.

Cross-wire netting.

Metal five-bar gate.

Electric fencing.

At the other end of the scale, at a fraction of the price, is a strained wire fence. This only requires posts at about 60–80 yard (55–70 m) intervals with lightweight posts in between, and consists of six or seven taut steel wires. It is cheap and effective requiring little mainten-ance, but nasty injuries can occur when horses get their feet hooked over the wire and there-fore it is not preferred. In fact it is not prudent to use wire in any fence below waist height for this reason, but where wire is used opt for plain wire as barbed wire can cause trouble-some scratches or worse. Avoid any form of netting at all costs for the aforementioned reason. A recent innovation is the use of stretched rubberised belting between stout posts. This is available in a variety of colours and widths and is extremely safe and non-injurious while providing a resilient boundary, rather like the ropes of a boxing ring. It has a long life and low maintenance costs but its success depends on it being kept taut.

Gates may be wooden or steel, the latter being lighter and usually adjustable for height and angle. Make sure there are two means of securing them. If you have a latch or hook, provide a chain and clip also as horses have been known to operate latches and hooks with amazing ease, especially if they are in good working order and your gate is properly hung, i.e. does not need to be lifted in order to open it. It is imperative that gates are maintained sound and gateways should be gravelled or sanded for winter use.

Water supplies to paddocks

The water supply is very important. Ensure that it is adequate and functioning correctly. Make sure that the stop tap is working should you require to turn it off in winter and that the ball-valve operates effectively. The ball-valve should also be protected from damage by horses. Keep the trough clean inside to pre-vent the water becoming tainted. Troughs sited under trees are particularly prone to contamination from decaying leaves. If the supply is a permanently sited trough make sure that it is secure and cannot be moved by animals. Ensure that the ground around it does not become muddy by laying some hard-core. Concrete is not advisable as this can

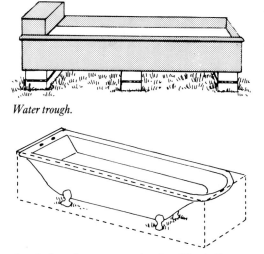

Water trough.

Iron bath used as water trough – dotted line indicates enclosed wooden boarding.

Parasite control

Horses at pasture cannot avoid picking up worms. It is therefore necessary to ensure that the risk is kept to a minimum by careful pasture management. The control of parasites in the horse is discussed in the veterinary chapter but it is appropriate here to mention the effect on the pasture. Ideally pasture should be rested from livestock for twelve months at a time but this is not practical in most cases. Worm eggs live in the horses' dung and are killed by sunlight. It is therefore a matter of spreading the dung by harrowing to expose the eggs to sunlight. This should be done as often as possible on heavily grazed grassland. This will also serve to aerate the grass and spread the dead grass, unclogging the roots of new seeds.

Where possible it is helpful to graze horses with cattle and sheep because their parasites are different to those in the horse and they help to break the worm cycle. It also makes grazing more consistent as horses are such bad grazers in that they prefer short grass and will ignore the longer grass which other stock will eat.

become slippery in frosty weather. A mobile trough can be used with a polythene pipe and moved to a fresh site when conditions dictate. An old bath is a useful alternative and easy to clean providing the plug is not lost. It should, however, be boarded around to safeguard the horse from injury.

CHAPTER FIFTEEN

CARE OF HORSES AT GRASS

Shelter and protection from flies

Shelter is an important factor both in winter and summer, from the wind and rain and from the heat and flies. Shelter can take many forms, from a high hedge to an open-fronted building facing south or south-east. A few trees planted in a corner of a field can be an invaluable asset. High hedges and lines of trees at right angles to prevailing winds can be effective as shelter for distances of five to ten times their height when foliated. Evergreens are therefore effective throughout the year.

Horses who are to be out at grass during the summer months should be protected from flies preferably by the provision of a field shelter. This will allow them to escape from the sun as well as from flies, which can torment them, make them very restless, and by causing them to roam about in an effort to avoid the insects, will make some animals lose condition. It must be remembered that droppings must be removed from the field shelter from time to time, not only for the welfare of the horse's feet but also to discourage flies. As a deterrent against flies harbouring in the field shelter, sawdust or sand can be sprinkled on the floor and sprayed with a disinfectant or proprietary anti-fly spray. The animals themselves can be sprayed with a fly repellent or if the horse is shy, spray a cloth before wiping the animal around the eyes, ears, muzzle and sheath. A group of horses usually afford each other protection from flies by swishing their

Field shelter

tails, often standing head to tail beside one another to achieve this. Competition horses who are turned out in the summer for an hour or so can be protected by a thin cotton sheet or New Zealand rug. The better bred and therefore thinner skinned the animal the more he will appreciate protection from flies, which can be made easier by turning out either first thing in the morning or in the evening. If the animal is to be turned out for longer it is often more comfortable for him in the warm summer months to be turned out at night and brought in to the shelter of a cool stable during the day time.

While trees will offer some protection to animals at grass there is always a danger of one being struck by lightning if standing underneath.

Turning out

If more than one horse is running in a field they should all have their hind shoes removed for safety. No matter how quiet they are normally their behaviour may differ when they are loose with other horses. Sooner or later a horse will be kicked and this can cause a blemish or worse. It is not recommended that mares and geldings be turned out in the same field unless absolutely necessary. In groups it will be found that horses often pair up and establish their own territory. This can often mean that the weaker animals will be bullied. If a horse is found to be a particular bully it is prudent to turn him out alone as it is not unknown for them to drive another out of a field, perhaps through a fence, causing serious problems. Whenever a horse has to be added to a group or taken away it is far safer to start from the stable, i.e. bring other horses in and then take them out together or to separate them bring them all in first. This will help eliminate any fighting, which often occurs, especially when horses have made friends. Whilst it is not a good policy to run a number of geldings with mares it is usually safe to run one gelding with several mares. (This goes back to their instinctive behaviour in a wild state of one male and his herd of females).

Likewise, it is better practice wherever possible to run horses of the same age groups together. If age groups are mixed the young animals are inevitably bullied by the older horses, particularly at feed times. Breeding mares should always be kept out of reach of geldings because any contact can cause the mare to slip her foal, particularly in the early stages of pregnancy. Mares and foals should always have a paddock to themselves but beware of overcrowding. Mares in the latter stages of pregnancy should be kept together and not mixed with barren mares or youngstock who may start galloping or kick.

Catching

It is a good habit to take a bowl or bucket with a few oats or nuts to encourage the horse or horses to come to you. If, however, there is a group of horses and only one handler it may be better to go without to avoid the animals crowding around you to compete for the food. Having said that it is not good practice to try singlehanded to catch one horse in a group. In fact it is a potentially hazardous exercise since the animals will often fight amongst themselves and the handler is in danger of being kicked or squashed. Often a few nuts or oats in a pocket will discreetly reward the one you want to catch without disturbing the others.

Extra care and subtlety will be needed to catch a shy or nervous horse. The headcollar should be kept behind the handler's back, out of sight, and once the animal is standing quietly the rope should be carefully passed over his neck until it can be reached from the near side and the animal realizes he is caught before the headcollar is put on. The golden rule for catching any horse, whether or not he is shy, is never make a grab at him even if he has a headcollar on. At all times he should be kept quiet and encouraged with the voice. Difficult horses are best left with their headcollar on whenever they are loose but with a short piece of string or cord attached to the underside will help the handler to catch him. It is often easier to let a difficult horse follow one that is easy to catch until he is in a building or enclosed space where he will be easy to catch. Headcollars, particularly leather ones will rub a horse if they are left on for any length of time, so they must fit correctly and be kept soft and supple. Another commandment to be remembered is never to herd horses in a round-up fashion. This will only upset them, make them wild and excitable and risk them kicking. Likewise a dog should never be allowed to chase them because the effect will be the same and the dog may also get kicked. The exception to this rule is of course when wild ponies and horses are rounded up on the moors and mountains by mounted riders. As with all forms of horsemanship care and attention to safety must be uppermost when catching horses. A gentle, quiet nature will encourage the horses to be calm while the slightest excitement arouses them and can cause a horse to become difficult.

A headcollar or halter should always be used to turn horses out and to catch them. The use of a bridle is very dangerous because even the quietest animal can get the bit caught in his mouth if he snatches to get loose which can result in the horse cantering away with the bridle hanging from his head. If the headcollar is to be left on, slip the end of the lead rope through the headcollar once the horse is in the field and unclip the rope, then, turn the horse towards you before slipping the rope out quietly without unsettling him in any way. The longer one takes to do this, the better stroking and talking to the horse to keep him calm and relaxed. Never slip the headcollar or rope in a hurry and encourage the horse to go away from you quickly because this will teach him to become impatient and bad-mannered. If there is more than one horse to be turned out at the same time be sure that each handler has control of his horse(s) and is at a safe distance from the others before turning them towards himself and releasing them at the same time. This avoids a horse snatching free with the headcollar and rope still attached to join the others. On no account should horses be released in the gateway. This is a most dangerous practice. They should be taken well out into the field with the gate safely closed.

Inspection of horses at grass

Whenever horse are turned away at grass they should be inspected at least once a day, but preferably both morning and evening. At the same time the water supply and fencing should be checked. The horses will need a close examination to establish the following points.

1. Ensure there are no wounds or swellings.
2. Check that the vital signs are normal, the nose and eyes are clear, the coat and membranes are normal, and there is no sign of lameness.
3. Watch for any abnormalities in behaviour and note them.
4. Look at the state of condition. If the horse has a long coat check that he is carrying enough condition by feeling his neck and ribs. Looks can be deceptive.
5. Remove any clothing to check that it is not rubbing anywhere and replace it. Straps and stitching should also be checked and anything suspect taken away for repair.
6. Look out for any excessive rubbing of the mane and tail.
7. In winter watch for mud fever, cracked heels or rain rash.
8. Feet will need to be trimmed regularly and watched for any cracks or soreness.

Feeding

Hay should either be fed in a rack or in a haynet tied high enough so that the horse cannot catch a foot in it when empty. Hay fed on the floor is particularly wasteful to horses who will tread on it and foul it. Concentrates should be fed in boxes or bins on the ground or mangers which hang over the rails or gate.

Outdoor hay rack.

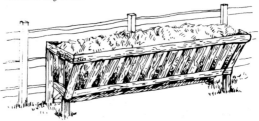

They will need to be placed well apart to prevent kicking and bullying. A spare manger will allow for horses changing places.

There may come a time, particularly with ponies, when the grass ration will have to be restricted, which can be difficult where animals are turned out in groups. Obviously some animals thrive better than others while age will also be significant. Good management and the eye of the horseman will be needed to recognise obesity and deal with it. Thoroughbreds rarely become overweight but hunters (espe-

Combined hay manger and feed trough.

Round hay manger.

cially the more common-bred ones) who are out for the summer can soon become too heavy. Excess fat causes unnecessary strains and stresses on the heart and limbs. If steps are not taken to restrict the diet the problems will also show when the horse comes back into work. Excess weight will take longer to work off and delay the fitness process. There is also the danger, especially in ponies, of laminitis ('fever of the feet') caused by being overweight. In their natural habitat ponies survive on vast acreages of poor quality pasture which in winter time can reduce them to a very poor condition which they then have to build up again in the summer. In contrast the home-bred, home-reared animal is generally kept in a consistently good condition the whole year round.

Permanent pasture is ideal for horses but fresh leys with a clover mixture can be too rich for grazing. Because of horses' very bad grazing habits it is preferable to have cattle or sheep to graze alongside horses. Small paddocks very soon become infested with worms and must therefore be rested from livestock

regularly and harrowed. The aftermath (i.e. the first growth of pasture following a cut of hay) can be rich and depending on the weather will be too rich for horses to graze *ad lib*. It may be necessary to restrict a horse's grass intake by turning him out either by day or by night but not both. If a yard with a shelter is available this would be ideal, especially for ponies, to save bringing them in to boxes which may otherwise be needed. If a pony is prone to being overweight it will be kinder to only allow him a few hours grazing per day.

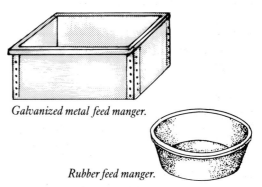

Galvanized metal feed manger.

Rubber feed manger.

Grass is at its most nutritious from the months of March to May when it is growing fresh, at which time the protein level and palatability is high. During this time a horse will tend to eat excessively but this will lessen as the grass becomes older and less palatable. Horses' bad grazing habits mean that they will graze in patches of short grass and leave apparently good grass which has become overgrown. The more limited the acreage, the more a horse will graze down to the soil in search of new fresh grass.

155

New Zealand rugs

It is often convenient to partially clip a horse or pony (say, with a trace clip) and have him live out with a New Zealand rug. It is, however, unreasonable not to provide adequate shelter even though the animal is rugged.

It is particularly important to ensure that a New Zealand rug fits correctly before turning a horse out because he has to be able to move about safely without the fear of it slipping especially after he rolls. Great care must be taken to ensure that it does not come loose or rub the horse sore so the correct size is imperative. New Zealands can be used for two things: firstly, in the turning out of a rugged horse for an hour or two in the daytime; secondly, as a permanent rug for animals who are to live out during the winter. As with any clothing there are varying qualities and it is a false economy to buy cheaply as such rugs will invariably not fit as well. The best is the original New Zealand rug which does not have a surcingle but, if it fits correctly, will right itself after the horse has rolled. It also has deeper sides and fits better round the neck as well as having soft chrome-leather leg straps.

Inferior New Zealands have a fitted surcingle which has to be adjusted regularly otherwise it will rub the horse's withers. The American cross-surcingle type again puts pressure on the withers and needs checking often. Many New Zealand rugs today are fitted with nylon straps which prevent chafing the legs. The fitting on the shoulders is equally important as the rug will rub as it slips back which it will do in no time if it is too small. Those horses who are permanently rugged with a New Zealand, will need to have it changed daily in wet weather for a dry one. When horses are wearing them all the time the rug should be checked daily to ensure it is not rubbing or that stitching or buckles are worn or broken. It is particularly dangerous for a leg strap to come undone.

When introducing a horse or pony to a New Zealand rug have him wear it in the stable or yard for a time to get used to the straps around his legs before turning him out in the field. Do not turn a horse out straight away until he is quite happy about the straps because he may panic and kick out at them without thinking where he is going and the result could be disastrous.

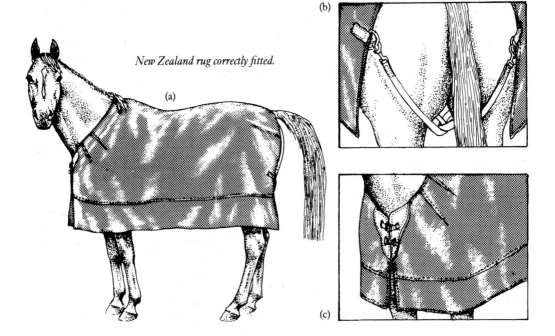

New Zealand rug correctly fitted.

(a)

(b)

(c)

A New Zealand rug will need to be well maintained for the obvious reasons of safety and as soon as it is not in use, say, in the spring, it should be soaked in a water trough or similar before being scrubbed with a yard brush and hose pipe. Repairs must be carried out, any leather straps oiled and buckles greased before the rug is put in store. A neglected rug will soon become hard with dried mud, stiff and uncomfortable for the horse apart from being difficult to handle when buckles and clips are stiff and rusty.

Turning horses away

Horses which have been stabled, and especially those that have been rugged, will need a roughing-off period to adjust to a change in lifestyle, i.e. climatic and environmental. This is particularly relevant in the spring when the weather can change drastically from day to day. A fine sunny day and preferably one when no night frost is forecast, is the time to turn them out completely. Once you have made the decision to turn a horse away do not be tempted to bring him in again unless the weather turns exceptionally bad (e.g. deep snow) because that is when a horse can easily catch a cold. It is rare for him to catch a cold through being turned away providing he has been roughed off appropriately. Roughing off is naturally easier and kinder in the spring than in the winter. A suitable roughing-off period is, say, two to three weeks depending on the type of horse and the weather prevailing at the time, with the horse going out in the day time and being in at night. At the same time the horse's rugs should be removed gradually and his diet adjusted, reducing the amount of hard concentrates. Any change in the diet should be a gradual process as explained in the feeding chapter and horses will have to be watched that scouring does not occur when they are introduced to lush, wet, spring grass. Hay should be offered during this time, but very often when horses have grass available for several hours they will eat very little hay, if any. This should not discourage offering it because the more the horse can be persuaded to alter his diet gradually the better it will be for his digestive system.

Rugs should be taken off one at a time over the roughing-off period until he is without for a few days before being turned away. He should not be groomed other than to ensure there is no mud under the withers and roller while a rug is still being used. This will allow nature to take over and encourage the horse to grow a coat and for the skin's natural grease to build up again to protect him from the weather. Top doors and windows should be left open and the horse's work programme adjusted so that he does no more than exercise at the walk. His shoes will need removing, unless the ground is very hard or there is a need for special shoes in front, and his feet trimmed.

Bringing up from grass

Again this process should be gradual, say over a week or so, preferably with the horse coming in for either the day or night to begin with. Some horses, especially youngsters, will be unsettled at the start of their confinement and will need to have the top door closed in case they push it open or at worse try to jump out again to join friends they may have been parted from. Once they have settled and accepted the change in routine, the top door and windows must be left open to allow for as much ventilation as possible. If the horse is stabled in a barn system put him in the end box nearest the door.

Once the horse is in permanently he will need to be wormed, his teeth checked and his feet shod. Some people choose to give a mild physic to clear out the digestive system and this is the time to do it. A word of caution here: with modern worming treatments and good management the use of physicking, as practised by previous generations, is less relevant today. The horse's condition according to the time of year will dictate the diet he will require. If the horse has been living off lush, wet grass he will clearly be softer in condition than one which has been grazing at the end of summer/beginning of autumn at which time the grass will normally be drier. This will make the transition to hay more natural. Generally speaking unless the horse is in very poor condition it will be sufficient to feed hay and a laxative meal in the evening to begin with until the horse begins to work, when his ration will have to be balanced with suitable concentrates. An even condition at the start, i.e. not too fat nor too thin, will make the early fitness preparation easier as the horse will not require extra miles of walking to remove excess fat. Clearly the horse which has been fed concentrates and hay in the field will be in a better condition than one which has been on a diet of grass alone.

Some important points to remember when bringing a horse up from grass are:

1. Ensure that his exercise is confined to the walk for the first one to two weeks depending on his condition. Do not ask for faster paces at this early stage.
2. Do not make the horse sweat unduly during this time.
3. Beware of the horse catching a chill if he gets damp, especially if he has a thick coat.
4. The horse's back and girth will take time to harden off to the use of saddles and rollers.
5. A roller left on in the stable will help to harden the area.
6. Do not try to canter until at least the third week.
7. There is no short cut to getting a horse fit.
8. A thin sheet or night rug will protect the horse from any draughts on his loins and help keep his coat flat and the dust off.
9. Beware in older horses of any filling in the hind joints.
10. Take care not to overfeed at the beginning of the fittening programme.
11. If a horse comes in from grass in poor condition and with a large stomach, it may be a case of worm infestation which will require the services of a vet who may need to use a stomach drench.

The use of surgical or methylated spirits will help to harden a horse's back when he first comes into work but care should be taken not to blister the sensitive skin of some horses.

Grass sickness

This is a disease whose cause is unknown although it is possibly viral. There is a frequent epidemic form found in Scotland. The symptoms are inflamed membranes, difficulty in swallowing, dribbling, depression, no bowel activity, a mucous discharge from the nostrils and foul-smelling breath. Food and water are often ejected through the nose.

The prognosis is most unfavourable because recovery is rare. Death is possible any time between twelve hours from onset to three days. In the chronic stages which may last for weeks or even months the abdomen tucks up and muscles become hard. There is no known treatment for this disease.

CHAPTER SIXTEEN

VETERINARY NOTES

The purpose of this chapter is to provide a broad understanding of the veterinary aspects of equine care. It does not dwell on the details of disease and injury, for which the reader has a choice of references, but rather it attempts to give practical guidance on the basic essentials of veterinary care so that every horse owner or groom will be able to carry out first aid. In many situations this will need a follow-up visit by the vet but as the owner's knowledge increases so too will his ability to expand on veterinary care. It must never be forgotten that common sense should prevail whenever veterinary care is required, especially in the case of an emergency when owners instinctively tend to panic unnecessarily.

Recognising when a horse is off colour

There are some symptoms or conditions which will immediately require the advice of a veterinary surgeon. These are most commonly: colic, azoturia, foaling or whenever the horse has an abnormal temperature. Often there are borderline cases when a horse may appear to be chronically ill but in fact is only displaying temporary signs of discomfort. Only experience will guide the horseman/woman in judging when to call the vet or when to observe the horse for a day or two to see if the symptoms subside.

If the horse comes out of the stable in the morning with lameness but no swelling or heat he may need the attention of the farrier rather than the vet to check the shoes and the feet. If, however, lameness persists, professional advice should be sought. At the first sign of heat or swelling in the tendon the vet should be called because the consequences can be serious and permanent. Many injuries occur within the confines of the stable and do not become evident until the horse is taken out. If there is any doubt as to the seriousness of any ailment, injury, disease or condition the vet should be called so as to avoid any delay in commencing treatment. It may be necessary to take the horse to the surgery if it is thought that large equipment such as an X-ray unit will be needed.

Signs of good and ill health

As stated in Chapter 2, the easiest way to remember what to look for in the horse is to remember the ABC – Appearance, Behaviour, Condition. A horse should have a bright outlook, prick his ears as he is approached and show an interest in his surroundings, bearing in mind that each horse reacts differently according to his character and temperament. Any abnormal behaviour such as restlessness, sweating unduly, laboured breathing, diarrhoea or straining should be considered early symptoms which may lead to other more positive signs of ill-health. The condition of the horse will depend on his state of fitness but first and foremost it must be healthy and not undernourished in any way. A horse in bad condition will show signs of malnutrition with ribs showing and hip and back bones protruding. He will have a lethargic attitude, his head hanging low and a general disinterest in his environment.

In the healthy horse the membranes of the nose, eyes and gums should be a salmon pink. Yellow membranes are a sign of jaundice. His eyes should be bright and alert. Running eyes are not necessarily abnormal but can be caused by a foreign body, flies or a reaction to touching something. Continual discharge may be associated with a cold. A nasal discharge of coloured mucus also indicates a cold or other more serious disease. Providing it is clear, a running nose is not essentially a symptom of ill-health.

How to take temperature, pulse and respiration rates

A veterinary thermometer should be shaken down to zero before the bulb end is inserted into the rectum at the side rather than centrally. This is because if there are any faeces in the rectum the thermometer will record a false temperature. A little Vaseline will make it easier to insert. Hold the tail high and to one side and hold the thermometer firmly in place for one minute. Do not let go during this time as it could be drawn into the rectum and disappear. On withdrawal do not touch the bulb end until the temperature is read. The normal temperature is between 99.8–101°F (about 38°C). Each horse's temperature varies slightly and will change under conditions of

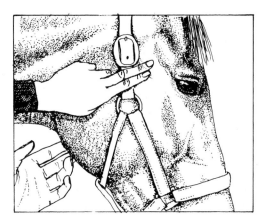

Pressure points.

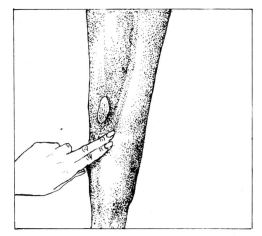

stress without necessarily indicating the horse is unwell. A rise in temperature to even 102°F is an indication that an unhealthy condition may be developing and the vet should be informed. The thermometer should be washed off in cold or tepid water after every use.

It is recommended that each horse's temperature under normal situations be recorded at various times so that it is easy to recognise when something is amiss. It is also useful in monitoring fitness to take the temperature immediately after exercise. If a temperature is taken before exercise and is found to be abnormal the horse must not go out. A slight rise in temperature is often indicative of the early stages of a viral infection. At the slightest sign of the horse being dull or off colour take

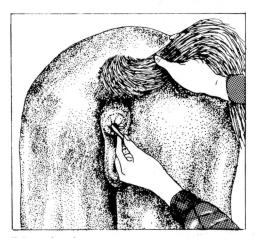

Taking a horse's temperature.

his temperature. Prompt response to a rise in temperature can lead to early diagnosis and treatment, often resulting in a quicker and fuller recovery.

The horse's pulse can be taken by placing two fingers either inside the jaw bone, high on the side of the head, or inside the horse's elbow. Only a light pressure is required to feel the throb of the pulse. A normal rate is between 36–42 beats per minute.

Respiration can be monitored by watching the horse's flank rise and fall. This is best done by standing immediately behind the horse, facing his quarters. At rest the number of breaths should be between 10–12 per minute.

TPR (temperature, pulse and respiration) rates will, of course, vary according to the situations in which the horse finds himself. Obvious increases will be apparent during and immediately after exertion and as a result of the many forms of stress which can affect him. The horse's temperament and type, i.e. cold or warm blood, are other contributing factors.

Nursing

A horse which is suspected of being in any way abnormal in his health should have his temperature taken four or five times throughout the day and more frequently into the evening too, if necessary. When a horse shows signs of being unwell he should be given small feeds at regular intervals, with the protein content reduced and all heating foods withdrawn, while being maintained on a laxative diet. In suitable weather conditions it may be beneficial in some cases to turn the horse out (with a New Zealand rug if necessary) to give him the chance to relax and graze. It is often the finest remedy when a horse is recovering from coughs, colds, azoturia or filled legs. It is important that the horse be kept warm at all times with rugs and bandages. Check that there are absolutely no draughts or leaks that could give the horse a chill. Again, if the weather is suitable, the horse can be fed outdoors providing he is warm enough. As a rule there is usually no reason why the horse cannot be groomed although it may not be

wise to strap him. If the horse cannot be turned out, lead him out three or four times a day to graze.

Owners should follow veterinary advice throughout a horse's illness and not be tempted to give the horse other drugs as they may well conflict with the vet's treatment. A reasonable time may be necessary for some treatments to become effective, depending on the form of administration, so one must avoid changing the medication out of impatience. 'Quack' medicines are used less today because it is usually found that modern veterinary treatments are more efficient. Medicines which have lost their label and instructions should be discarded because they may have expired their sell/use by date. Intravenous injections should only be carried out by a vet because a mistake can be fatal. Subcutaneous and intra-muscular injections, on the other hand, can be performed by a capable person who has been taught by a veterinary surgeon. Antibiotics are available for oral administration under the vet's instructions.

Medicine chest

A medicine chest should contain the following basic items for the purpose of administering first aid:

Gamgee tissue
Cotton wool
Stretch bandages
Surgical scissors
A non-stick dressing
Animalinex
Kaolin poultice
Thermometer
Disinfectant
Antiseptic spray
Antiseptic wound powder
Variety of bandages
Epsom salts (up to 8 oz (225 g) a day can be given as a laxative in the feed or in water)
Colic drench
Gauze dressing
Lint
Fly repellent
Iodine ointment or lotion

Liniments for both bony and bursular leg
swellings
Vaseline
Magnesium sulphate paste
Antibiotic aerosol
Antibiotic cream

Vaccinations

The ideal time to inoculate a horse is when he
is off work because he should not be worked or
travelled for a couple of days after. Vaccinations are available and indeed required by
many equestrian bodies, against equine
influenza and tetanus. It is possible to combine
the two and top up with an annual booster.
Vaccination certificates will need to be maintained and carried for competition horses
where evidence of vaccination is required by
the sport's ruling body. Certificates must be
kept current and signed by the veterinary
surgeon. If a horse changes hands the certificate should accompany him.

The teeth

The horse's teeth will need examining by a
veterinary surgeon or horse dentist twice a
year. Some indications that the horse is in
need of dental treatment are: evasion of the bit
whilst being ridden, food not properly masticated; quidding; and playing with food. The
groom should be in the habit of looking in the
horse's mouth every day to check that the bit is
not causing any soreness, that teeth are in
good order (especially in youngstock) and that
gums and lips are not injured or sore. Regular
attention will also help the head-shy horse to
overcome any apprehension. He should be
handled with quiet patience because once he
becomes difficult around his head and mouth
he will often develop a permanent problem.
The person who knows the horse best is most
suited to handling him for the vet or dentist
because it is hoped that the horse will have
developed confidence in him/her between the
vet's/dentist's visit. If necessary a bale or stool
can be used for the groom to stand on, with
due consideration for safety.

Wolf teeth will need to be checked from
time to time to ensure that they do not become
troublesome as they develop. Some remain
small and do not cause the horse any discomfort while others become bigger and pointed,
interfere with bitting and like the wisdom
tooth in humans, can be painful. The vet or
dentist can usually pull them out with forceps,
sometimes with the help of a local anaesthetic.
Because they are small compared to molars or
incisors they can be awkward to get hold of
and care must be taken to ensure that the root
is not left in. Often wolf teeth are shed in the
young horse but many remain and need
removing artificially. It is said that only male
horses have wolf teeth but occasionally mares
may develop them.

Molars need to be rasped once or twice a
year to prevent any uneven sharp edges
developing. Some vets or dentists prefer to
rasp the teeth without any aids while others
prefer to use a gag which prevents the horse
from closing his mouth and enables the dentist
to work unheeded. Neglect of tooth care can
lead to numerous problems, the most drastic
of which is malnutrition. Sharp edges on the
molars can cut into the horse's gum and cause
bleeding if ignored. The parts which tend to
wear unevenly are the outer edge of the upper
jaw and the inner edge of the lower jaw. These
can be felt by the groom by taking the tongue
out to one side while carefully examining the
jaw on the opposite side.

The incisors are not usually any trouble
once the young horse has shed his milk teeth.
It is wise to keep an eye on him during this
process to ensure that there are no complica-
tions but, generally speaking, nature takes care
of this process. Once the horse has a full
mouth the incisors will alter with age and only
an accident interferes with them.

A parrot mouth is one where the top jaw
overlaps the lower and is usually looked for
when purchasing a horse because it can, in
extreme cases, impede the horse's grazing.
This is because the top and bottom jaws do not
meet evenly at the incisors. This is also referred to as being 'over-shot'.

Faeces or droppings

The state of the horse's droppings is an indication of how the horse's digestion is working, his diet and general well-being. A horse under stress will often pass loose droppings as he would if he had diarrhoea. Worm infections and stomach disorders can also cause loose or watery droppings. Constipation will result in hard, round balls of dung. The norm is droppings that are easy for the horse to pass without him showing any signs of discomfort, and which break as they hit the ground. There should be no foul odour, sliminess or odd colour.

Urine

The horse should be able to pass urine comfortably about four to five times a day. The urine will usually emit a strong smell of ammonia when it is fresh. The colour varies from a pale yellow to a dark brown depending on the horse's diet and condition. It can also appear clear, or thick, which is often the case with mares. If the horse strains himself to urinate this is an indication that his kidneys are not functioning efficiently and he should be treated on veterinary advice.

Detecting lameness

Unless there is heat or swelling in a particular limb it is often difficult to accurately detect and diagnose lameness in the horse. The animal's conformation and action can sometimes be misleading in detecting the seat of lameness. In many cases it may be easy to locate heat and/or swelling in the tendon, foot or joint and subsequently straightforward to make a diagnosis, especially if the area is tender to touch. However, a thorough examination should always be carried out and no conclusions drawn prematurely. Experience is the greatest asset to any horseman, particularly regarding the horse's soundness. Often a lameness is suspected as being caused by a sprained tendon when in fact it is due to a poisoned foot. It is prudent therefore to examine the horse most carefully, first by walking him in hand on a hard level surface (as explained in Chapter 2) on a loose rein so that his head can move freely and any nodding can be easily seen. (A nod of the head implies lameness in a front leg and will be witnessed when the good leg hits the ground.) He should then be trotted up in the same way, after which he can be turned in a small circle on either rein. It is important not to allow the horse to trot too fast as this may disguise any lameness. The horse can be reined back and then turned about to test for any spinal injuries or stringhalt. The horse should be watched carefully in his stable to see how he stands and moves about and if he favours a leg and how. For example, pointing of a fore leg may denote corns or laminitis, whereas holding a fore leg behind the normal vertical position is associated with ligament or shoulder injury.

It is quite normal for a horse to stand with a hind leg resting and this does not necessarily reflect an unsoundness. If the vet is called to diagnose a lameness it is vitally important for him to know the horse's full history of the injuries or lameness to help him to arrive at his diagnosis. The groom or owner should therefore record the details and relate them accurately to the vet. Once the vet is consulted the owner must resist the opportunity to suggest his/her own conclusions which will only confuse the vet, but rather confine his remarks to the symptoms and treatments and the outcomes relevant.

The management of the lame horse must be effected promptly with emphasis on his diet and exercise. The degree of lameness is reflected in each pace, i.e. it is common to become more lame at one pace than at another. In the case of a splint lameness the horse may appear sound at the walk but not at the trot. Navicular, on the other hand, improves the soundness with exercise, often referred to as 'warming to work'. Spavin lameness in the hock is usually continuous. A muscle injury often makes the horse trail the foot of the affected limb. Sprains and strains of ligaments and joints are generally accompanied by heat and inflammation. In all cases it is imperative to the immediate and long-term soundness of the horse that he is not, under any circumstances, allowed to return to work until all signs of swelling and heat have completely disappeared.

Lameness of the hind leg is often more difficult to diagnose. A vast number of lamenesses in the hind leg are located in the hock area. By turning the horse in a small circle or trotting on the turn a lameness can often be pin-pointed.

The feet should always be examined closely, especially if the horse has recently been attended by the farrier and may be shod with new shoes. The foot is often the seat of injury or disease. Commonly, when there is pus under the sole of the foot, a swelling over the flexor tendons is evident.

Filled legs

Filled legs are swollen limbs which can be accompanied by heat caused either by injury or disease. Strains affect the fore legs more commonly than the hind legs, often one leg at a time. Legs can become filled if the horse is unfit or as a result of a digestive disorder, in which case a laxative diet may be necessary as a cure. New oats or hay may cause filled legs. Sometimes only the hind legs are affected. Filled legs can also be a symptom of a viral infection and will need to be monitored by checking the horse's temperature. In this case the horse will need resting. Some horses, particularly in their advancing years, may have slightly filled legs which are sometimes referred to as 'gummy', caused by years of work and strain. There may be no treatment required if the horse remains sound but stable bandages fitted overnight will usually reduce the filling. In many cases the swelling in the legs will often go down with exercise. In all cases it is important to keep the swelling soft. Swellings which are neglected may become hard and difficult to cure.

Detecting heat

It should become a habit to run a hand down the horse's legs regularly especially when a horse is in work. This way the owner or groom can become familiar with each individual horse's legs and learn what is normal for that animal. Each horse's circulation is different and some horses' legs may appear warmer than others. It is up to the handler to accustom himself so that any abnormality is recognised early and dealt with promptly. Failure to identify and diagnose at an early stage will delay treatment and may prolong inactivity.

Lameness or swelling is often accompanied by heat which must be diagnosed so that if there is any infection, the pus is drawn out by the application of poultices. The modern practice is to allow an infection to take its course and any poisonous substances encouraged to mature rather than be suppressed by the use of antibiotics. It is felt that by masking the infection with antibiotics, the actual seat of the infection is not always located. If there is heat but no swelling the owner should satisfy himself that the cause is not in the foot by testing with hoof-testers. This can be done by the vet or farrier who will then investigate the sole of the hoof with a paring knife if the horse is tender in that area. It may be that the infection is in a development stage which can take several days to mature with the help of poultices, tubbing, hosing and fomenting. If on the other hand there is no indication of infection in the foot but the swelling is in the leg a sprain must be suspected. In this case the horse must be rested and given a laxative diet.

Many accidents and diseases can be prevented by thorough and careful management, by recognising symptoms and knowing how to treat them promptly. Faulty conformation, working young horses for too long, unsuitable exercise, working unfit horses, bad feeding and just plain carelessness are common causes of many problems which beset the horse. The horse's soundness relies on accurate diagnosis and treatment but most importantly knowing when to rest a horse and save him from a day's work when there is perhaps only the slightest indication of something amiss. It is better to be safe than sorry and miss perhaps one day's sport rather than jeopardise the horse's career.

Colic

Colic is an illness similar to indigestion in the human. It is potentially very serious and in severe cases, life-threatening. The vast majority of colics are, however, not serious and can be treated very simply by injection. The drugs available today are very effective although their one drawback is that they are so effective they can mask the pain caused by very severe colic, such as a twisted gut. The symptoms of colic are severe stomach pains, which can lead to a twisted gut in extreme cases. The horse may kick at his stomach, look back at his sides, sweat and get up and down often pawing at the bedding in a general uneasy state. Horses have a very low pain threshold for abdominal pain: what would kill a horse would probably not even make a cow uncomfortable, which is an indication of the enormous difference in their sensitivity. One of the reasons for checking stabled horses late at night is to ensure they are not restless because of a discomfort which may be caused by colic. Any unusual noises in the stable like banging or kicking should be investigated and not ignored, because this is an indication of the horse's general unease. There are many causes of colic, the most common of which are dietary i.e. a disturbed feeding regime; excessive water intake following exercise or food; sand off the bottom of a stream, in soil or by a water tank (this is referred to as sand colic); fast work too soon after a feed; worms; or something which has fermented in the intestine and produced an abnormal amount of gas. Sand colic, which is the result of horses continually licking water near sand, causes scouring which may lead to a weight loss and can be fatal. Wind-suckers and crib-biters are also prone to bouts of colic.

Colic can be classified as follows. Spasmodic colic is when a length of intestine goes into spasm and clamps down causing a great deal of pain and irritating the gut wall. Flatulent colic is when a build-up of gas is caught in the gut wall. Impaction or a stoppage in the alimentary canal is caused by a blockage of food. A horse which has eaten his bedding can cause an obstruction in his large bowel of very dry material which will not pass through quickly. Worms can also cause this build-up as they are being expelled after worming. A large percentage of colic cases are caused by worms. Damage to the blood supply for the intestines can be caused by red worm infestation. Blood-worms, too, can cause an artery blockage by thrombosis. It is not uncommon for a horse to suffer more than one attack of colic in his life. A sudden change in diet or moving a horse to richer pasture are other causes which form a stoppage. Horses which bolt their food may also develop these same symptoms.

At the first sign of colic the vet should be summoned. Never leave the horse unattended; preferably two people should stay with him because if he thrashes about in pain someone could get injured. Safety must not be overlooked if the horse is reeling about and the handler is trying to get the horse to his feet. It must be remembered that a horse is very sensitive to any change in diet or to the effects of stress and that the intestines are literally a long tube of muscle which can become fatigued. It is quite common for colic to develop after exercise particularly if the horse has been fed just before it. When this happens the food may be in the upper part of the intestine and the stretching effect of the food is inhibited because blood is diverted away from that area to the muscles and to the lungs in order to cope with the exercise that the horse is doing. This results in problems caused by reduced activity in the small intestine. The muscle becomes fatigued and goes into spasm. In these cases treatment to encourage the intestine to move again and make the passage of its contents easier is required. This is when a bland type of oil is useful. An ammonia and morphine mixture will reduce the sensitivity of the intestine and make it less painful. A laxative effect is desired to allow food to pass through more easily and give the muscles time to relax without there being a large volume of content to stretch them. Enemas are also used for a large bowel obstruction because they stimulate gut movement. A stomach tube will enable the vet to administer oils and drenches which would otherwise be difficult or impossible to get the horse to swallow.

There are many colic drenches available and it is easy to mix one's own concoction. A drench of linseed oil or liquid paraffin will do no harm to the horse and will not interfere with whatever treatment the vet may prescribe. If the horse is obviously in some discomfort it does not necessarily alleviate his suffering to walk him around endlessly. Providing he is kept warm and quiet he will be just as well off in the stable. If, on the other hand, he is in danger of injuring himself by reeling about he would be better off in an indoor school or paddock. A large horse in pain, throwing himself about, is sometimes best sedated before he injures himself but that is up to a veterinary surgeon.

Post-nursing care is equally important and the horse must be given time to recover fully before he is put to work again. It is important to remember that the intestines are a muscle which have been stressed by colic and need time to recover with the help of a laxative diet and easy exercise for a few days. Establishing the cause of a colic attack will help in treating it and prevent further cases.

Wounds

These can vary from a simple cut or puncture to a major laceration. Whatever the size or severity it must be remembered that any wound can become infected if it is not treated properly and promptly. If there is any doubt as to whether the wound can heal itself the vet should be called to decide if stitches are necessary. Even a couple of stitches in a small cut will prevent proud flesh and an ugly scar. If the vet has been summoned the wound should be washed in water only and a clean dressing applied. The use of preparatory treatments at this stage may interfere with the vet's own treatment. Beware of granulated tissue (proud flesh) forming on an open wound because it is difficult to remove once it is established. Minor wounds should have the whole area cleaned and, if necessary, any long coat trimmed to prevent dust and hair entering the wound. Many superficial wounds and grazes

will respond to being cleaned and an antiseptic spray or powder applied. Anything more serious should not be left to heal itself but the vet called to advise.

Puncture wounds are caused by an object entering the flesh and making a hole. Often the site is particularly vulnerable to infection as the puncture will be caused by a sharp object which will not be clean and may even be rusty. Blackthorns can be to blame as well as rusty nails, glass, wood or even a fork. Poulticing helps to draw out any infection but the wound must be kept clean and the vet may be needed to administer antibiotics. Puncture wounds in the foot can prove very troublesome and care will be needed to treat them and ensure that the sole of the foot does not become under-run causing a recurring lameness. Drainage is often difficult so it is important to ensure that the wound is cut out well to encourage the infection to drain. If this is not done thoroughly the drainage point may block up and contain the infection causing it to exit elsewhere, perhaps at the coronet or frog. This may take a long time and cause undue suffering to the horse.

Whatever the nature of the wound be sure that the horse is vaccinated against tetanus. Early treatment will encourage a quick and complete recovery from a wound. Carelessness and neglect can result in permanent scarring, proud flesh, thickening of the skin and, if the wound is infected, a possibility of recurrence. Where an infection is present there is a danger of a fever developing, indicated by a rise in temperature which must be monitored and the vet called accordingly.

Coughs and colds

The first symptoms of a cold are when the horse loses his appetite, appears lethargic with a dull eye and may begin to cough. The coat may become stary, the eyes and nose runny, the breathing increased, and the horse may shiver or show signs of breaking out into a cold sweat after exercise or travelling. His temperature should be taken at frequent intervals, e.g. five or six times a day, and at the first sign of any abnormality work must be stopped. There is a danger of permanent damage to the horse's respiratory system, as well as to his heart from the stress involved, if he is worked while suffering from a cold or cough. The winter time or periods of damp and cold with extreme changes of temperature are the most likely times when a horse can catch a cold. Also, when he returns from long periods of work, such as a day's hunting, or when he does not dry off thoroughly after sweating, especially when he has a winter coat.

He must be allowed as much fresh air as possible without draughts as long as he is warm enough. Providing he is checked regularly to ensure that he is warm, in some cases the horse is as well in a paddock, rugged up, where he is encouraged to put his head down which will allow his nasal passages to drain freely. When he is stabled he can be fed on the floor for the same reason. All feed and hay must be damped to discourage dust particles from aggravating him. Haylage is often preferred to hay and the horse may be bedded on shavings or paper rather than straw which will also irritate the horse's nasal passages.

A nasal discharge will accompany a cold and this can vary from a thin light-coloured mucus to a thick yellow mucus which can become lumpy in chronic cases. The vet will be required to attend the horse and administer a course of antibiotics to relieve the condition as soon as possible. There are a number of cures and remedies available on the open market for coughs and colds but the vet is the best person to advise you, for some treatments may suit one horse and not another. A cough or cold is highly infectious and will quickly spread around a yard, even if the horse is isolated. At all times the horse must be carefully nursed and kept warm. There is a risk of a cold leading to pneumonia if the horse is not cared for properly.

There are different types of cough ranging from a short sharp one to a deep, rough, persistent, dry cough and all will need to be treated individually under veterinary supervision. The owner should be advised as to what work/exercise if any the horse can carry out safely. Coughs can often be extremely stubborn to cure and the owner must not be impatient to return the horse to normal work while there is still evidence of a cough as this will risk permanent damage to the respiratory system. Where a runny nose is accompanied by swelling of the jaw and glands this can indicate a mild attack of strangles which is a highly infectious disease quite different from a common cold. The swellings will eventually develop abscesses which will emit pus (yellowish viscous matter produced from inflamed or infected tissue). In this case the vet must be called immediately and the horse isolated.

An anti-flu vaccination is compulsory for horses participating in some sports, such as racing, show jumping and eventing, but it does not necessarily preclude the horse from contracting some strains of virus. If the horse becomes sick with symptoms of inappetite, a dull outlook and in some cases filled legs, and there are no other obvious symptoms, this can indicate that he has caught a touch of a virus. Often large establishments can become infected and take some time to get rid of the virus. In the meantime the horse must be confined, the vet called and exercise limited. Some strains of virus are as yet not clearly identified and are consequently difficult to cure. The vet may take blood samples to examine in an attempt to identify the strain. Often a cough will accompany a viral infection making it particularly contagious. Foals and yearlings are susceptible to a viral infection known as rhinitis, which can cause a high temperature, give the animal a sore throat and make it difficult for him to eat, although he may not necessarily have a cough. Nursing must be thorough if the horse is to respond to treatment within, say, a couple of weeks, and aftercare is equally relevant for a complete recovery.

Horses, especially youngsters, coming in for the first time after a spell outdoors will often catch a cold more easily than those being turned out, so if possible they should be isolated for a period to acclimatise.

Drugs

A list of forbidden substances is compiled by the Jockey Club and issued through the Racecourse Security Services at Newmarket. It includes everything from aspirins to oxygen. The criterion is a non-normal nutrient, i.e. anything which is not present normally is considered to be a prohibited substance, and the list therefore includes all drugs and in some cases substances which are present normally but which if they exceed normal levels are considered abnormal. The means of detecting forbidden substances is now so sophisticated that it is possible to detect the presence of a few parts per million. The most common drugs that affect the majority are the anti-inflammatory and non-steroid treatments which overcome pain that would otherwise prevent a horse competing.

In eventing the permitted level of Butazolidine is 4 micrograms per millilitre. If a horse is on Butazolidine it is difficult to calculate the level of intake as his blood level is not constant but influenced by diet, the rate of absorption, the rate of metabolism, breakdown and excretion of the substance, the horse's general state of health, his activity and so on. Therefore if a number of horses are on the same dosage the results of blood tests are likely to be different for each horse. Basically the principle is not to permit pain-killing drugs which allow the horse to work when he would otherwise not be able to do so.

In dressage Butazolidine is not allowed at any level. Show jumping allows 5 micrograms per millilitre of blood plasma. The game of polo does not lay down any rules on the use of drugs probably because, unlike the aforementioned disciplines, it does not involve prize-money.

Sedatives and antibiotics can leave traces of stimulants which may show up if the horse is dope-tested within, say, forty-eight hours. Random dope tests are taken regularly and the penalty for offenders is a strong enough deterrent to dissuade them trying again.

The following paragraph is given as advice for users of Phenylbutazone:

The Royal College of Veterinary Surgeons has given guidance to the veterinary profession regarding the elimination time of various anti-inflammatory drugs and they state 'if a veterinarian recommends the discontinuance of any such drug not less than eight days before racing (even though a period may be longer than is necessary in many instances) he should be able to feel sure that he has catered for all but the most exceptional case. . . . Important exceptions to this advice, however, are various anabolic steroids and certain other drugs which are specifically formulated to have a sustained effect. It is strictly recommended that when such drugs have been used they should be discontinued at least forty-two days before racing.' The Royal College make it clear when offering this advice that they 'cannot accept responsibility for the possibility that a drug, though not likely to fall within the category covered by the paragraph above, might nevertheless behave in a similar manner. Nor indeed can responsibility be accepted for any atypical horse which may take longer than normal to clear a drug.'

Any drug will take at least four to five days to work through a horse but vets usually recommend that users are allowed seven days to be clear of its effect.

Tranquillisers are often prescribed for use on nervous or vicious horses which have to undergo veterinary treatment and are also useful for clipping horses (as previously mentioned in Chapter 10). They can also help to travel a nervous horse but of course this would not be permitted for competing. Phenylbutazone as such is bitter and complicated to administer but more often nowadays Equipalazone is used which is 'Bute' coated to make it more acceptable to the horse. This is effective as an anti-inflammatory for acute and chronic muscular injuries, laminitis and other similar lamenesses which affect bones and soft tissues. One of the side effects of 'Bute' is that it dries the horse's saliva and makes him insensitive in the mouth. There are other brands available which must only be used under veterinary guidance. It is a misconception to think that by multiplying the dosage, soundness will be achieved sooner. Remember that pain is a symptom of lameness which drugs only serve to disguise while helping to reduce inflammation. While anti-inflammatory drugs play a useful part in reducing inflammation they should not be abused by owners seeking to mask a condition which may need other treatment.

Drugs are not always easy to administer orally and are best fed in the evening meal which has been mixed thoroughly. If the horse objects, a small feed of succulent food, say crushed oats, sugarbeet, perhaps with carrots and apples, can be mixed and offered to the horse before he has his normal ration.

The use of anabolic steroids in horses must be done under strict veterinary supervision. There is a lot of abuse with steroids, as with 'Bute', for the benefit of short-term gain at the horse's expense. Steroids act to build up the muscle in such a way as to make the horse look in an impressive condition. But as soon as the steroids are withdrawn the animal is likely to drop away in condition. There are side-effects which affect mares and can make some geldings unmanageable in company. Steroids are sometimes used for horses who are entered in a sale to make them look more impressive. The subsequent waste of condition afterwards is alarming to the new owner.

Certain substances in the horse's feedstuffs are forbidden under some rules, particularly racing. It is therefore prudent to have feed materials inspected to be sure of what they contain, bearing in mind that each feed batch is variable.

Dope-testing is carried out by taking samples of saliva, urine or blood. For taking urine samples the horse is put in a stable with a straw bedding and left with an attendant until the horse stales. If after an hour the horse has not volunteered to stale a blood sample is usually taken instead. If an owner refuses to allow his horse to undergo a test, he is invariably banned from future competitions at the discretion of the ruling body or breed society.

Nose bleeding/Epistaxis

Idiopathic spontaneous pulmonary haemorrhage is the full name for this condition and comes from the lungs and not from the nose. The reason appears to be mechanical, the result of a wedge of lung becoming compressed between the diaphragm and backbone. Blood vessels in the dorsal diaphramatic lobe of the lungs rupture, resulting in free blood in the airways which is blown up and excreted through one or both nostrils. In severe cases the horse can drown from being unable to get enough air into his lungs due to the air spaces being full of blood.

The exact cause is unknown but it is believed that respiratory viruses and stable dust will exacerbate it. Stress and pressure of hard work usually found in racing are most commonly associated with it. In most cases it does not cause any great threat to life but obviously restricts the horse's ability to race because of the limit of oxygen he can get into his lungs to fuel his metabolic processes. It is often found that the horse will become weak at the end of a race, cough and discharge blood. In other cases a horse may perform disappointingly and although not show any immediate symptoms will have a nose bleed a couple of hours later. Generally bleeding stops fairly soon of its own accord because the vessels are small, unlike an artery, and no treatment is necessary. In excessive cases coagulants based on oxalic acid can be given. Vitamin K is also useful and there are other commercial products available which reduce haemorrhage and encourage blood clotting. Once a horse suffers from Epistaxis, prevention from a recurrence is recommended by avoiding stress and ensuring that the environment is as dust-free as possible.

Another condition which causes the nose to bleed is found in the guttural pouches, which are two areas at the back of the jaw bone wherein lie the internal carotid artery on each side and various large nerves. The entrance into that cavity is from the back of the pharynx and from here fungal infections sometimes enter. If an infection establishes itself on the wall of the carotid artery and erodes and weakens it, the result can be a sudden massive haemorrhage. The horse can quickly bleed to death. Treatment involves ligating the blood vessel to stop the haemorrhage.

It is often found that a horse may have a slight nose bleed at any time without any apparent reasons because it is not associated with stress. Endoscopic examination into the guttural pouch will enable the vet to determine whether there is any fungal growth on the outside of the blood vessel wall, which would weaken it. If there is, the vet can then tie it off immediately.

Minor nose bleeds can also occur at any time as a result of a polyp or similar growth in the nostril.

Excessive bleeding

With a small wound it is recommended that a tourniquet is immediately applied. It must be tight enough to stop the bleeding but should be removed every ten minutes to allow the blood to reach the cut-off area, making sure that the tourniquet is re-applied with enough pressure. If the area is small it is helpful to apply something like a pebble to give added pressure. In the meantime the vet should be called.

In the case of excessive bleeding to a large area a bandage or similar material which is absorbent should be applied, which will form a sort of seal. As the bleeding continues additional bandages should be fitted on top of the existing ones. Under no circumstances should the soiled bandages be removed before another is applied. The horse should be kept as quiet as possible. Any movement will increase his heart rate and therefore his blood pressure, forcing blood out of the wound. Horses can lose an enormous amount of blood, even as much as a bucketful, before being in great danger. Once the bleeding has stopped it is very unlikely that it will start again unless very large blood vessels have been cut. In the event of a rider being stranded in open country, say out hunting, providing the wound is stuffed with a stock or handkerchief for the

blood to clot into and there is a house nearby it is better to move the horse very gently to shelter rather than risk him catching a cold by standing still. Remember that blood is full of fibres and will clot itself to form a seal. If bleeding is profuse the horse may need other treatment, perhaps a general anaesthetic and stitching up of the wound so he will need to be moved to a building anyway.

Every situation is different but the danger is always that the owner/rider will panic when a horse is bleeding. The first thing to remember is to remain calm, keep the horse quiet and take him indoors where possible so that further attention can be given and the vet called. Common sense is very important but unfortunately it is not very common.

Common treatments

Common treatments which will be used in most yards at some time include poulticing, fomenting, tubbing and hosing.

Poultices. These are a means of reducing swelling and bringing out bruising. Bran poultices are used mainly for poulticing the hoof by soaking a quantity of bran in hot water and applying it with either a poultice boot or plastic bag over which a sack and bandage is fitted to hold it in place for twelve to twenty-four hours. Care must be taken to ensure that the bran is at blood temperature before applying it otherwise it can scald the skin. A poultice to the foot will need to be secured very well to prevent the horse getting it off. Animalintex is a popular proprietary poultice which can be used for most instances which require a poultice. Again it will need soaking in hot water and wringing out before being applied and secured with a stable bandage. The advantage of Animalintex is that once it is removed from the injury it is possible to see on the dressing if any infection or foreign body has been drawn out of the wound. Kaolin is used as a poultice dressing as a safeguard against swelling after hard work or when there is inflammation without an open wound. This is heated in a tin and spread on brown paper then left to cool slightly before being applied. To retain the heat plastic

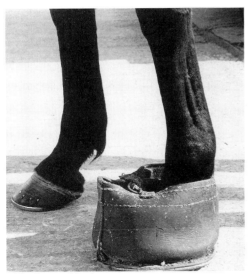

A leather poultice boot.

through the sole overnight. The equine boot which is designed to replace a shoe and guard against the hoof breaking up is not practical as a poultice boot.

Tubbing is when a limb is soaked in cold or warm water to reduce swelling. Often the two water temperatures are alternated to give a hot then cold effect in an attempt to reduce inflammation, and draw out any infection. Standing the horse in a stream is an effective way of cold tubbing to cool the tendons and help reduce swelling.

Hosing with a hose pipe is one of the more common means of cooling the limbs and reducing inflammation. A horse can be hosed for any length of time and it is often effective to alternate with some hot treatment such as poulticing, as in the case of tubbing.

Fomenting means applying warm, damp flannels to an inflamed part of the horse in order to sooth and reduce the bruising or infection. Once applied they can be covered with a waterproof sheet or dry blanket and changed from time to time so that they do not lose their heat.

is fitted between the paper and bandage. Again, care must be taken not to apply the poultice too hot and risk scalding the skin.

A custom-made poultice boot is very useful particularly in a large yard. Although it may not be used very often and is expensive to buy it is very handy. Gamgee tissue and a bandage fitted around the leg first will guard against marking and rubbing the tendon sore. A poultice boot will need maintaining in a clean and supple state, being washed, dried and oiled thoroughly after each use otherwise it will smell and the leather will dry and crack. When Animalintex is used it will need to be secured by a bandage before fitting the poultice boot. An old-fashioned lawn boot, designed to be fitted to a horse who had to pull machinery on a grass surface to prevent him from marking it, is an alternative to a poultice boot although difficult to find today. Rubber poultice boots are least expensive and becoming more common but they are not so hard-wearing as the leather type.

A corner of a hessian or plastic sack serves as an alternative to the poultice boot but this must be fitted well with a bandage to keep the poultice in place. Elastoplast bandage, although expensive, is well worth using around the foot to keep the poultice securely in place and prevent the horse from wearing

Alternative treatments

There are a lot of 'alternative' treatments available today, some of which have been borrowed from human medicine but only a few, as yet, have been proven with horses.

Ultra-sonic treatment is fairly well established for treating sprained tendons and ligaments or softening areas of fluid swellings before they are drained. Care must be taken that it is not used at too high a frequency or intensity or for too long as this can have a detrimental effect on tissues, similar to microwaving. It is safer to use the equipment at a lower frequency than is commonly practised. Capped hocks and elbows may also benefit from ultra-sound by softening the swelling and thereby making it more easily absorbed.

Electro-magnetism. There are a range of electro-magnetic machines on the market which have originated from use in human medicine where the treatment was used on fractured bones which did not knit together.

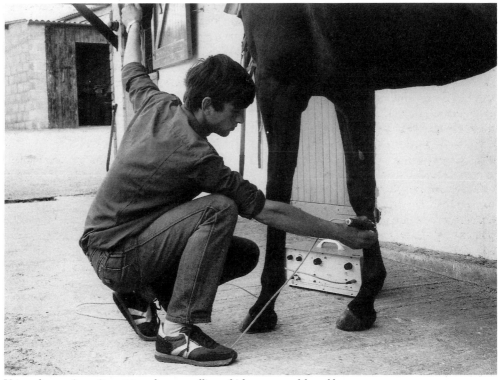

Using ultra-sonic equipment to reduce a swelling which was caused by a blow.

Occasionally fractured bone ends heal over but, during this healing process, the ends stay separate and do not stick together. This is known as a non-union fracture. By passing a pulsed electro-magnetic field across the ends of the broken bone, a better and quicker healing response was achieved. It is therefore logical to assume that the same thing would happen to other damaged tissues although recent research shows that it must be a pulsed magnetic field or a moving magnetic force (rather than the magnetic field from a magnetic which is non moving). To be effective an alternating pulse magnetic wave is needed. There are boots which have a static magnetic field but these do not seem to be as effective. **Lasers** are similar in that they are part of the non-visible spectrum of electro-magnetic waves and do appear to have a healing effect. The treatment involves high-frequency light vibration although the lasers used for horses are known as cold lasers with a fairly low

frequency. Positive results have been achieved using lasers on sore shins and tendons for example, but again this treatment is still relatively new and perhaps its value is not fully appreciated. Laser treatment can be used daily for seven to ten days but it is safe to say that if treatment is successful it will be evident after just a few days. If a response is not obvious then there is usually no point in continuing the treatment.

There is, of course, the possibility that these alternative treatments may be used for too many different ailments and in the course of time lose their credibility. It must be stressed that physiotherapeutic machines should only be used by skilled operators and not fall into the hands of laymen. In professional hands with a safe intensity applied there are generally very few problems. Having said this, like so many treatments there are dangers. They can, for example, cause burning and tissue necrosis (death of tissue, especially bone) through bombardment by pulsed waves.

Firing is a method of applying heat therapy to the skin with the use of hot irons to cause an inflammatory reaction. The normal process of repair to damaged tissue which the blood supply effects by inflammation is in some cases inadequate. Firing will repeat this process in a more exaggerated way.

Some typical conditions for which firing is used are soft tissue damage, spavins, osselets, splints, chronic arthritis and sesamoiditis.

There are two forms of firing, one is 'line firing' when the hot iron makes lines across the skin or 'pin firing' (sometimes known as point firing). The latter method causes a series of points to the skin usually in a rectangular or diagonal pattern.

Firing has become increasingly controversial because many people believe it is inhumane and barbaric. The degree to which a limb is fired can, however, be fairly well controlled. Some horses become demented after they have been heavily fired or blistered and in their case the treatment is no doubt barbaric, but the majority of horses do not resent firing as much as has been claimed. The discomfort can be very largely overcome and managed within tolerable limits. The main problem is that at the moment there does not appear to be any scientific rationale to firing. Advocates of this procedure are still unable to offer an intelligible, substantiated opinion as to what firing actually achieves. The arguments in favour of firing seem to be purely based on circumstantial and subjective evidence as a result of an individual's interpretation founded on historical experience. It is very difficult to find sufficient clinical evidence to examine the effects of firing. Also, the basis of each injury is different, the standards of management, care, treatment and exercise after injury vary from horse to horse. It is no easy task to establish something which can be used as a basis to compare different forms of treatment. The only evidence which supports firing is that burning any skin will stimulate a healing response generally throughout the body. Having said that there

are much better means of limiting the destructive changes that occur in tendons after injury than waiting and firing because this will stir up the inflammatory process. An inflammatory process is very often useful in healing but if it is uncontrolled it is destructive. A self-digestive effect can result as happens with meat that is hung whereby tissue enzymes leak out of damaged cells causing further damage to the remaining cells.

The most effective time for treating damage is during the acute stage, i.e. immediately after injury when the greatest benefit and improvement can be achieved. This immediate treatment and subsequent good physiotherapy are the key factors in treating most injuries.

Carbon-fibre implants. This is another technique which has been borrowed from human medicine where it is used in the repair of small tendons and ligaments. By replacing damaged tendons with plaited strands of carbon-fibre the tendon cells actually reproduce and grow along the line of the carbon-fibre forming new tissue laid down along the line that the tendon is working. Scar tissue generally tends to be laid down at random so, for example, in a horse's tendon which has been left to heal itself the scar tissue fibres can be across the line of the tendon and in any direction, making it weaker. Much of the tendon's power is in the intrinsic strength of the tissue and also in the way in which the tendon is constructed and the fibres aligned. It therefore follows that by educating the repair process of new tissue to grow in the same direction as the line of stress it should be stronger. The problem in horses is the size of the tendon. Carbon-fibre implants are difficult to insert without causing a lot of tissue damage during the operation.

Originally the tendon sheath was cut open and the carbon-fibre inlaid but because of the massive reaction and adhesion between the tendon and surrounding tissue this method has largely been superseded. The alternative method now practised is to slit the tendon in three places – the top, middle and bottom – and then thread the fibre in. Carbon-fibre tends to break down and spread all over the body and can be found in the lymph glands, so nowadays man-made fibres are more often being used instead. Having said this, fibre implants are going out of fashion slightly because of the heavy unsightly thickening which results. All tendon treatments can claim, quite justifiably, some fairly dramatic results but there is no one treatment that is consistently successful. There have been as many disasters with each type of treatment as there have been successes.

A badly damaged tendon is likely to take something in the region of eighteen months to heal fully and properly after initial treatment and physiotherapy. The blood supply in a tendon is quite poor and the number of cells (which reduce with age) that produce tendon fibres are quite low which makes the healing process very gradual. This tissue has a very slow metabolic turnover making their healing very slow. The weakest part of the tendon, subject to the greatest risk from injury and indeed the most common site, is the middle third because it has the poorest blood supply. The tendon sheath comes from the knee above and the fetlock below but in the middle it runs outwards making this part most vulnerable. Owners must therefore resign themselves to allowing a lot of time, patience and, of course, money in the treatment of tendons.

Emergency treatments

First of all call the vet and explain to him exactly what has happened to the horse. The most important thing in an emergency is not to panic. Owners invariably become excited, ring the vet every five minutes and cause a general state of pandemonium. Often when the vet arrives he has very little to do and all the panic and excitement has been unfounded. The situation is never as dramatic as it appears to the owner.

The horse should be kept quiet and avoid anything which is likely to distress it. If necessary move him somewhere where he can be examined in reasonable comfort and good light. Ensure there is plenty of good light, clean water and that the place is clean. If there is an injury, it can be bandaged; as long as it is a

fairly clean dressing it will not be harmful. There is little more the owner can do while waiting for the vet apart from keeping the horse quiet and warm. He can be offered a haynet, rather than a feed, to encourage him to chew slowly and take his mind off the pain. Often a horse with a serious injury such as a bad attack of colic, or one dying of laminitis, will try and eat as a reflex distraction to whatever is causing the pain. This will help the horse to feel less distressed about what has happened to him. Unless the horse is likely to need a general anaesthetic it will not do any harm to give him a haynet.

Use of antibiotics

Fewer antibiotics are used on horses than other animals because most injuries and problems which affect horses are physical injuries caused by managerial problems. Recognising early signs of injury are terribly important as has been previously mentioned and this is where the owner's knowledge and experience are fully recognised. The range of antibiotics used in horses is quite narrow compared to other species. Straight penicillin is perhaps the most widely used antibiotic in horses. Potentiated sulphonamides are also very useful for respiratory and gastro-intestinal problems.

Parasite control

One cannot prevent horses picking up worms. Worms have evolved with the horse as a parasite. Parasites by definition do not kill their hosts otherwise they kill themselves. If the host is dead they have nowhere else to go because the adult form cannot survive outside the horse. The egg and larvae forms have a limited life span on pasture although some eggs can survive even through the winter. The problems arise because horses are confined to small fields, crowding the horse and parasite together. It is therefore possible for a horse to develop an abnormally high and life-threatening burden of parasites.

There is a wide range of parasites which affect horses. Some like the ascarids affect foals and yearlings, but adult horses are less vulnerable to them. Tapeworms are not very common in horses and don't seem to cause many problems although, of course, the horse is better off without them. The most problematical worms are the strongyles because they migrate through the tissues. The worst parasite of all is the redworm, of which the *Strongyles vulgaris* is the most dangerous. At the larval stage when they are ingested they burrow through the intestinal wall and travel up the blood vessels to the mesentary artery where they cause thrombosis and colic. As a result of recent research it is now believed that colic and the main damage to the system is caused by the mass migration of these worms through the intestinal wall so it occurs at an earlier stage in the larval migration than was originally thought.

Strongyles are peculiar in that they have a period of dormancy where the larval worms are eaten off the pasture, they burrow into the intestine or intestinal wall. Some stay in the intestine and others migrate outside and then become dormant for several months. They will then emerge and return to the intestines. Once they have returned to the intestines to complete their life cycle they then start to mature and reproduce and lay eggs. It takes approximately six weeks after they have returned to the intestines to lay eggs in the summer time, a little bit longer in the winter. If the horse is wormed inside that six-week period the worms are killed before they are mature and start to reproduce. This is why the recommended worming interval is four to six weeks, particularly if they are picking up fresh worms and the whole cycle is continuous. If worming is left longer the female worms will have reached maturity and are capable of laying thousands of eggs each per day. In a badly infested horse with a heavy worm burden, millions of eggs could be produced each day which will very quickly contaminate the pasture.

Worming should therefore take place every six weeks, or eight weeks at the most in winter.

Bots should also be treated for in the spring and autumn. Eqvalan Ivermectin is particularly effective against migrating larvae. Other products have to be used in massive doses to kill off larvae outside the intestines. There is a range of drugs which are derivatives of Thiabenzole. The trade name of worming products gives no indication of which drug it contains. However it is helpful to understand that some drugs are Benzimadazoles, such as Equizole and Panacur, while the most common non-Benzimadazole is Strongid. To avoid a resistance developing against worm treatment one can alternate between, say, Panacur and Strongid although it is not necessary to do this too often. One could use Strongid most times alternating with Eqvalan in the spring and autumn. There is, however, no reason why Eqvalan cannot be used throughout, but it is more expensive. Resistance to worm treatments which is what one is trying to avoid is passed on from one generation of parasites to another. Providing one can kill a population of worms at the generation interval, the next generation is effectively prevented from appearing. Also if the drug is changed at that interval, the next generation of worms is exposed to a drug which works in a completely different way. It is therefore not necessary to change in less than a nine-month period because most serious parasites have a life cycle of about nine months. Also it is not necessary to change more often because this gives the worms a chance to build up a resistance to a number of drugs, effectively defeating the object. It is, however, recommended that veterinary advice be sought regarding parasite control and a worming programme to suit each horse. The vet will also guide you on pasture management, which is, after all, the origin of the parasites.

Back injuries

It appears to be fashionable at this time for the horse to have a back problem and much money is made from the so-called treatment of back injuries. Manipulators can at best reduce muscle spasm, which is very painful and very inhibiting. On the other hand, in many cases they do no good at all because they cannot effectively cure serious back problems only superficial back problems. If the superficial problem is a muscle spasm which is the chronic result of an injury, manipulators can do some good. If, however, that injury has settled down with time but has left the horse with a muscle spasm and the original injury was not treated successfully then the spasm is likely to return. This is when often the manipulator will continue to return to treat the same horse.

The horse's back is very strong and the bones do not come out of place. What actually happens is that the soft tissue around them can become disturbed but the vertebrae themselves do not move otherwise one would have a paraplegic horse! Bad back injuries usually result from a fall and massive bruising to spinal ligaments and spinal muscles, and even a fracture of the spinal processes. Some athletic horses, such as show jumpers and steeplechasers, can impinge the action of the dorsal processes of the vertebrae. This causes a low-grade pain resulting in back muscle spasm which is set up by the horse's own reflex system to limit the amount of movement in the painful area. Typically, if the horse has a real back problem he is abnormally stiff in his back and may carry his tail out tautly to one side or behind him. He will also have a reduced hind limb action and hind leg impulsion with a very short choppy stride behind. In most cases, however, such horses do not become lame.

Often horses are referred for a second opinion because they have a stiff back when this is due to some lameness and not a bad back. Stifle and spavin lamenesses are sometimes wrongly diagnosed as a back problem because the horse is stiff when in fact he would be sore in his back anyway as a result of a prolonged lameness. Therefore the first thing to do when back pain is suspected is to eliminate all other possibilities, i.e. lamenesses, stifle injury and so on, before examining the back as a final option. While soreness in the back may be more common as a result of ill-fitting saddles or clothing, actual back damage is rare.

The one place where a horse's back can move is his pelvis which can become rotated out of line. From the withers to the pelvis there is almost no movement between the vertebrae.

Blood testing

Routine blood tests are normally only necessary in racehorses where their progress needs to be monitored because of the problems which arise from viruses in racing yards. In most cases blood testing will only need carrying out where it is suspected that a horse is not thriving or that it has not performed as well as it should. It is a very helpful means of investigating a problem rather than looking for one. The presence of infection and anaemia are the two most common results which the vet would look for in a blood test. A horse can be tested for worms via a blood sample but in most cases a good worming programme will save the need to spend that money. It is, however, useful to measure the protein level which will give an indication as to whether or not there are any migrating strongyles outside the intestines which the normal worming programme would not kill off.

Contagious diseases

At the first sign of coughing and a nasal discharge, an abnormal temperature or any other indication of a contagious disease the animal/s concerned should be isolated with their own groom. Great care must be taken to ensure that no clothing (human or equine), tack, grooming kit or other equipment comes into contact with healthy horses because disease will spread through contamination or even on the wind.

Ringworm is a highly infectious disease which contaminates materials as well as animals and persons. It is identified by either large, round, crusty mounds or patches of raised hair. Cattle are often a source of infection and can spread the disease through materials such as fencing. Iodine can be painted on the patches of skin or a proprietary spray

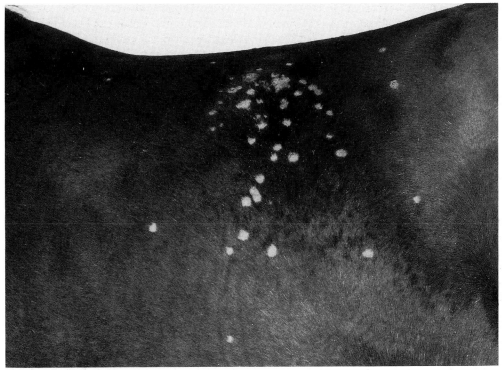

An example of ringworm.

used. A vet may give powders or tablets which can be administered in the feed. These will often cure the infectious spores which are not visible.

Canadian Ringworm or Pox is one of the most severe and difficult contagious diseases to cure. The early symptoms of small circles around the girth area are often mistaken for sweat rash. Unless it is diagnosed and treated early it will soon infect the whole yard. Any raised hairs or bald patches should be recognised and dealt with promptly before an infection spreads, as contagious diseases can be transmitted in many ways. The handler should himself beware of irritant or inflamed skin because many of the contagious diseases which horses contract can infect humans and be carried by them. The horse's skin may become hairless or with scaling skin or have the appearance of a rash. It is highly irritating and causes the horse to rub himself against anything solid thereby infecting whatever he contacts. Because of its highly contagious

nature it must be treated promptly to arrest the spread. The horse's clothing should be confiscated once the hose is rid of the disease and the stable thoroughly scrubbed, preferably steam washed. Ideally the woodwork should be creosoted before the stable is used for another horse. All equipment, e.g. grooming kit, tack, haynet, feed bucket etc., that has been in contact with the horse must be washed in disinfectant or washing soda. Re-infection can occur several months after the initial outbreak, which is an indication of how highly infectious this disease is and how scrupulous one must be to contain and eradicate it promptly. Vehicles used for transporting horses can also carry the disease and will need hosing down, preferably with a pressure hose. Vets and farriers are also in danger of spreading the disease and should therefore be asked to change their outer clothing before entering your yard. A disinfectant foot dip at the yard entrance is a useful safeguard.

It is for such diseases that the facility of an

177

isolation box is essential to segregate the patient. (The provision of an isolation box is explained in Chapter 4.) Horses stabled in integral blocks, such as in the barn system, are most vulnerable to infectious diseases and it is difficult to control the spread. Once a horse is suspected of having an infectious disease he must be isolated as far away as possible from the other horses until he is safely cured and there is absolutely no evidence of any symptoms remaining.

Treatment aids

Cradle. A cradle may be needed to prevent a horse from touching his front legs or interfering with a dressing. Fitted to his neck it prohibits the horse from turning his head and is therefore particularly useful when a horse has received treatment to the fore legs, such as blistering or firing when he could blister his nose. It also prevents him interfering with bandages, undoing stitches, etc.

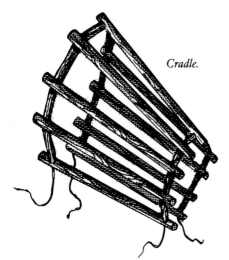

Cradle.

When the horse is first fitted with a cradle he should be watched for a little while until he has settled in case he takes fright of it. The stable should be checked to ensure that there are no protrusions on which the cradle could get caught. It may be wise to close the top door of the stable to prevent the horse from catching the cradle on the bottom door.

Racking up also prevents the horse from touching himself providing he is tied short enough and cannot harm himself. A strong headcollar should be used and preferably two rack chains fitted as cross-ties. If a horse is tied for any length of time he should be given a haynet regularly to keep him occupied. A sod of turf, carrots, apples or similar appetisers will amuse him and help relieve the boredom.

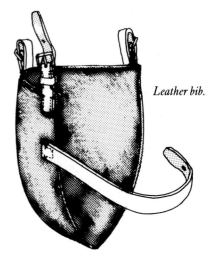

Leather bib.

Bib. Fitting a bib to the back of the headcollar will usually prevent the horse from reaching his rugs and front bandages but it must be deep enough and tied tightly to the headcollar otherwise the horse will play with it and eventually find a way around it. There are plastic and leather ones on the market but if these prove to be too shallow and the horse can evade them, a home-made bib might be the answer. The top of a plastic or rubber bucket can be cut to the right shape and holes punched in the rim for string to pass through. The bib can then be tied to the headcollar and will prove a useful, made-to-measure alternative. **Whirlpool boots** are becoming increasingly popular. Although the price is rather prohibitive they are often justifiable in large establishments where there are a number of horses in work. A whirlpool boot is similar to a large wellington boot, in which the horse's leg is placed. The boot is then filled with cold water and a small motor attached circulates the

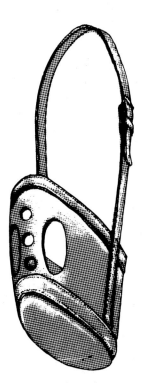

Leather muzzle.

water. Horses generally find this acceptable once they become used to it because the effect is soothing and the motor is of low voltage and therefore relatively quiet.

Most conditions which affect the legs and require cold water treatment will benefit from the use of whirlpool boots once or twice a day for as long as recommended by a veterinary surgeon.

Muzzle. Made either of wire mesh, leather or plastic a muzzle can be fitted to the horse without a headcollar to prevent him from eating. It will not, however, stop him from drinking out of a bucket but if his stable is fitted with an automatic bowl he will need a water bucket instead. The muzzle is useful not only for preventing the horse from eating if he is to be starved but also from eating his bed. It will also stop him from interfering with clothing or crib-biting. It should only be used as a temporary method until an alternative is found because it must be unpleasant for the horse to wear for a long-term. Some old types of leather muzzle will prevent the horse from drinking and cannot therefore be left on for any length of time.

CHAPTER SEVENTEEN

OTHER ASPECTS OF MANAGEMENT

Types of yard

Riding yards can be categorised according to the horse's purpose but principally they are either private or for commercial use. Private yards are those where the owner keeps horses or ponies for his own pleasure and does not offer horses and equestrian-related services or facilities to the general public for hire or reward. Commercial yards are defined as those which provide a service to the public, e.g. riding schools which hire out horses and/ or teach students who are training for a career with horses. They can also be specialist yards which train horse and rider for a specific discipline.

The following breakdown is representative of the types of yard which operate on a private and/or commercial basis:

Competition or training – private or commercial – e.g. eventing, showing, show jumping, polo, driving, dressage, long distance riding, western riding, hunting, heavy horses, tie and ride, gymkhana, pony club, triathlon, Arab racing, side-saddle equitation, point-to-point and general hacking.

Livery – commercial – for any purpose.

Hunt kennels – run by the hunt, funded primarily by subscribers.

Dealing – commercial – where the horses or ponies are bought and sold as a business.

Trekking centres – commercial – riding-holiday establishments which provide ponies for trekking.

Racing – commercial – these are specialist yards producing horses for flat and National Hunt racing, run by trainers who hold a licence or permit respectively.

Riding school – commercial – an establishment which operates under the Riding Establishment Act, that is with a licence from the Local Environment Office, and hires out horses to the public, teaches riding and prepares career students for a qualification. These yards may have the approval or recognition from the British Horse Society and/or the Association of British Riding Schools.

Stud – private and commercial – Thoroughbred and non-Thoroughbred breeding.

In addition there are multi-purpose yards: for example, a livery/riding school: TB stud/non TB stud; or dealing/competition yard.

Livery yards

A livery can vary according to the owner's requirements and pocket, from a full livery which caters for all the horse's needs in the owner's absence to the DIY type of accommodation which allows for the owner to look after the horse himself and just rent the stable and facilities. Specialist yards are those which take in horses for a particular discipline or purpose such as for breaking, schooling, eventing, show jumping, showing etc. and hunting.

A breaking yard generally works on the basis of preparing a horse for the owner to ride away within a given period of, say, eight weeks, according to the age and condition of the horse as well as the level of his education at the outset. A completely unbroken youngster will need mouthing, lungeing, long-reining and backing before he can be ridden away in company and in traffic up to the stage when he is safe enough for most experienced riders.

In many cases (say, with a three-year-old or under-developed four-year-old), a young horse may not be strong enough to take much work at this stage and so only the minimum amount of his education would be established before he was returned to his owner and turned away to mature. It is only fair on the horse that the owner ensures that the horse is in a good enough condition to start work and not waste anyone's time by expecting a livery yard to produce a horse that is under-nourished and therefore weak as this may lead to problems of injury or illness. The horse must not be expected to begin work unless he is healthy and in a thriving condition. There have been barbaric cases of cruelty to young horses where the animals have been weakened to quieten them. This is a deplorable practice and causes unnecessary suffering. A capable horseman will not need to revert to such callous methods in order to control and discipline a young horse. These old-fashioned methods will often scar a horse mentally for life so be sure that the breaking yard you choose is a reputable establishment with a good record of producing young horses in a calm and efficient manner. This will only be born from experience in the natural horseman and so considering that this early education is the foundation of the horse's career it is wise to take every precaution.

Very often the proprietor of the livery yard may not necessarily be the person who will break your young horse so it is up to the owner to find out who will and to satisfy themself of their ability.

The specialist yards will normally accommodate a horse for their particular purpose for any length of time that the horse is in work and if they have the land they may be able to turn him out whilst he is on holiday. The owner may have chosen the yard on its reputation or for convenience. If for the former he may expect either the owner to school his horse personally or a capable member of staff under the proprietor's supervision. He may also pay for the professional rider to help him at com-

petitions too and for lessons at home which, of course, will inflate the livery bill.

Show horses and ponies are often produced by a specialist yard which will also provide the jockey if necessary. These establishments rely on a successful record in competitions which is, after all, their shop window and they are therefore very professional in every way.

Racing yards, that is permit and licence holders for the flat and NH racing, take horses for as long as required on a full livery basis whereby the horse's needs are taken care of from his general welfare to training and racing. Trainer's fees vary from yard to yard as to what each considers a basic livery and what is regarded as extra. For example, keep and training may be taken as basic and the following as extras: blacksmith – i.e. shoeing for exercise, plating for racing; veterinary care – from surgeons' charges to veterinary requisites; gallop rent per horse on either private or public gallops; transport and travelling expenses at home and to overseas meetings; lads' expenses to races; trainers' expenses to races; gratuities to stable staff depending on the value of the race; percentage of fee for retained jockey and work riders; commission on sales and purchases; entering for races and riding charges via Wetherby's; racing colours; and sundries, which may cover items such as postage, telephone and telex.

Most racehorse trainers undertake to advise the owner with regard to breeding and purchasing or selling with many representing the owner often anonymously. The relationship between the owner and his trainer is crucial to the success of his horses. A close liaison is essential to discuss the horse's racing schedule. It is, however, a matter of individual negotiation and personal preference as to how much the owner wishes to be involved. Some owners try to influence the trainer when he would prefer to use his own judgement. A trainer once said, 'It is not so much training the horses but the owners.'

A half livery is when the owner pays for his horse to be stabled, fed, mucked out and sometimes groomed, plus the use of the yard's facilities such as a manège and paddock, while the owner takes care of all the horse's exercise and work requirements. Each yard varies as to how much responsibility falls on the owner and what the stable staff will carry out on the owner's behalf.

A grass livery is when a horse is kept at grass continually and the owner negotiates with the proprietor how much attention he would like his horse to have. This may involve the horse being inspected once or twice a day, checking his feet and arranging for the blacksmith to attend, calling the vet where necessary and mostly acting on the owner's behalf. As with a half livery the owner may wish to carry out some or most of the chores himself and each yard will have its own rules.

*A DIY livery is when the stable and perhaps the yard's facilities, such as manège or paddock, are rented while the owner takes care of the horse. This is becoming increasingly popular, especially in suburban areas and with the one-horse owner. He will also be expected to provide all the horse's needs, such as bedding, fodder, mucking out, feeding, grooming, exercise, arranging for the blacksmith and vet. It is often possible to purchase all the feedstuffs etc. from the livery yard and to liaise with the proprietor over travelling the horse and fitting the blacksmith's visits in with the other horses' requirements.

With most liveries it is possible to come to some arrangement over travelling the horse. Some establishments have their own transport which they are happy to hire out while others will hire transport on the owner's behalf and ensure that the horse is transported to a competition and accompanied by a groom if required.

Hunting liveries are sometimes essential where the hunt requires that the horse is stabled within the hunt country. The yard usually provides all that the horse may need from when he is brought into work in August or September, depending on whether he is going cubbing or if he is to be ready for the opening meet, right through the season to when the horse is roughed off again, usually the end of April/beginning of May. All the fitness preparation, clipping and general wel-

fare is taken care of and the horse prepared for a day's hunting. On hunting days the horse will be prepared, i.e. plaited up, tacked up and transported to the meet, where the owner is waiting. The owner is later met and the horse taken home again after hunting as part of the livery service. Many hunting owners will see their horse only on hunting days; the rest of the time he is the complete responsibility of the livery yard proprietor.

Some owners prefer to have more involvement and may exercise the horse. Some also may have their own horse transport with which to convey the horse from the yard to the meet. If the owner chooses he may call upon the livery yard to arrange the horse's grass livery whilst the horse is off work. The livery yard will usually need the owner to provide him with saddlery and clothing for each horse but maintenance and repair can be dealt with by the proprietor if required. It is up to the owner to ensure that the horse is fully insured against loss of use and anything else he chooses such as veterinary fees, saddlery, transporting and fire or theft as the livery owner's insurance policy will not necessarily cover all of these items.

Before sending a horse to livery it is up to the owner to negotiate with the proprietor so that each understands fully and exactly what is expected of each party. The owner must satisfy himself as to what services and facilities are available and their arrangements, for example, booking the use of a manège or arranging for the horse to be turned out; who will turn out the horse and fetch him in?; who will groom the horse and how much is he expected to do?; what is regarded as basic livery and what items will be considered as extra? If the horse is to go into a specialist yard for competitions the owner will need to discuss the aims and ambitions for his horse with the trainer. He needs to establish that the trainer is happy about the horse's potential, otherwise the owner may become disillusioned and frustrated apart from wasting time and money.

There are some aspects of keeping a horse at livery which are regarded as etiquette. The owner would be wise to observe them for the sake of all concerned. Always let the livery yard know when you plan to visit, especially if you intend to ride and if you need the horse prepared for you. If you want to arrive un-announced because you are unhappy about the horse's welfare that is the time to move the horse to another yard. Once trust is broken it will never be repaired. Never forget the groom and acknowledge his work by a 'present' from time to time, especially at Christmas. It is discourteous to the proprietor for the owner to discuss his horse with the groom first. If you are happy that the establishment provides the service and facilities you require, be content to pay a fair market price. To bargain can result in the horse suffering from lack of attention and possibly even food.

Exercise and training facilities

In recent years facilities for schooling horses have become increasingly sophisticated. An outdoor manège or indoor school for hire can often be found within reach for those who do not have their own. More and more private yards are now equipped with an all-weather surface on which to work.

Racing yards need gallops for daily use and public gallops can be rented if the trainer does not have his own. Gallops can be either on turf or a prepared all-weather surface topped with bark or sand. Grass gallops have to be maintained by mowing in the spring and summer, and harrowing and rolling has to be carried out all the year round. However, once the area has been designated and farmed maintenance is relatively inexpensive. At the most it will require re-seeding every few years and alternative gallops will be needed whilst the new grass is establishing itself as the gallops cannot be used for the first twelve months.

By comparison an all-weather gallop is more expensive to install and maintain. An area of top soil has to be removed before the new surface can be laid. The material is costly to buy because a large quantity is needed to cover the surface area and establish a suitable depth. If the material is laid on top of the ground it will scatter with use so a wide enough gulley will need to be dug out so that the finished surface lies just below the surrounding land. Once the gallop has settled in it will need rolling and/or harrowing regularly, preferably daily, to ensure that the surface is consistent, free from hazards and the correct depth (it may need topping up). If any wet patches develop a drainage site must be built otherwise a horse may harm himself by hitting a boggy area when travelling at speed. At worst he could fall and unseat his rider.

An outdoor manège is costly to install because the foundations need careful preparation. The success of the working surface will depend almost entirely on the foundations and drainage so there is no point in spending a lot of money laying the surface if it does not drain efficiently. This can put the manège out of use when it is most needed. Where a membrane is used (i.e. a plastic layer between the foundations and the surface) it will need to be laid deep enough so as not to be dug up by the horses with continual use. This is a common fault.

The area chosen must be level, preferably free-draining and fenced off. It will need easy access for large vehicles but, if possible, should not be sited too near passing traffic. The more quiet and secluded, the easier it will be to gain the horse's concentration. The manufacturers of the surface material will advise on the construction. First decide on your requirements, i.e. how much the riding area is to be used and which type of surface will cope with the wear without having to be watered to lay the dust or salted to prevent freezing. It will pay to shop around and not just settle for the product which is marketed the best. What will suit one yard may not necessarily suit another. The best test is to ride on different surfaces and jump if necessary to test for slipping. It is also prudent to establish the success of each company's surfaces to satisfy yourself that they have stood up to wear and tear over a period of years and not just months.

The most common all-weather riding surfaces are woodchips, sand, rubber and PVC. Woodchips or woodfibre are suitable for all purposes and drain easily. Both soft and hard woods are used, sometimes mixed together, but each one is practical. They can also be compressed or blended with rubber or bark. Sand was one of the first materials used for riding surfaces but has been found to create problems with bad drainage, freezing as well as dust. It requires little maintenance, is a good surface for jumping but does not lend itself well for dressage because of being deep and heavy. Rubber is very hard wearing and can comprise shredded waste rubber such as car tyres. The latest form is a nylon-fibre reinforced rubber matting. All rubber-based materials offer good drainage and ride well providing the foundation materials have been carefully selected. PVC, in the form of chopped cable or plastic granules (from which any metal has been removed) is, like rubber, hard wearing, drains very well and does not need much maintenance. It stands up to jumping probably as well as any material and is virtually indestructible.

All surfaces depend on the preparation of the foundation and subsequent regular maintenance. The size will vary as to requirements but the regular size is 40 m × 20 m (130 ft × 65 ft) although dressage specialists will often need 60 m × 40 m (195 ft × 130 ft). The arena edges will need to be boarded, high enough to contain the material, and the surface raked in as it is spread outwards by the horse. The horse will take out enough by the hoof-ful without losing any unnecessarily.

In essence the owner must look for a good riding surface which has proved hard wearing and requires as little maintenance as possible. The advantage of having a riding surface which is serviceable for 365 days a year is worth the expense to the competition yard which cannot afford for its horses to be out of work for any length of time.

While a manège will offer the rider a consistent working surface he must not be tempted to ride the horse exclusively within this confined space as it will lead to boredom and staleness. At all times, especially with a young animal, the horse must be ridden correctly; that is to say when working he should go on the bit with the rider ensuring that the horse is properly balanced, particularly on the corners. Harm can be done to limbs, especially joints, if the horse is not ridden correctly in a confined space and allowed to go in a lazy manner. Although finance will be the limiting factor when choosing an arena it should be remembered that the bigger the better. Young horses, particularly, will suffer from being asked to work in a confined space before they are sufficiently schooled and therefore balanced to cope with tight corners. They can be damaged mentally and physically.

An indoor school will call for the same preparation and construction in its surface with the added consideration of siting a large building. Basically the area will be much the same, although the foundations will involve more work to provide for concreting uprights etc. Again, the construction company will take care of everything from initial site inspection to the finished product.

However the main complication with an indoor arena, as opposed to an outdoor manège, is that building permission is required from the local council. This applies also to stabling and any permanent building, with the exception of field shelters. Difficulties of obtaining building permission arise in urban areas where a large building may be an eyesore to local residents. The more rural the situation the less likely the threat of opposition. It is illegal to erect a building in England without first obtaining building permission. Failure to do so will result in the owner having to dismantle it and possibly face a fine as well as the delay of starting over again through the proper channels.

Rates and rating

Stables, schools and equestrian ancillary buildings, such as jumping arenas and outdoor manèges, are all ratable. Farms, however, are regarded differently which is why stud farms, at the time of writing, are seeking the same status. Stables which are near to a house are likely to be rated with it but if they are away from the main residence they may be rated separately.

Legal requirements

There are a number of Parliamentary acts which affect stables, the horses and stable staff. All of these are available from Her Majesty's Stationery Office and the British Horse Society. English law comprises three principles: Statute Law – Acts of Parliament; Case Law – law of precedent; and Principles of Equity – guiding principles.

The *Riding Establishment Act 1964 and 1970* covers the UK except Northern Ireland and applies to stables which provide a service to the public by hiring out horses for reward. Annual inspections are carried out to consider the condition of the horses, i.e. whether they are healthy and fit for the job; the suitability of stabling and paddocks; the horses' welfare, e.g. food, water, exercise, rest, bedding and grooming; veterinary provisions; health and safety matters (at least one person must be trained in first-aid); fire precautions and suitability of ancillary buildings. If a horse is found to be unfit for work he must have a veterinary certificate before beginning work again. It is illegal to hide a horse from the inspector or to use horses under the age of four years. The person in charge or responsible for supervising horses hired out must be over fifteen years of age. An insurance policy to cover employers and public liability is another legal requirement. Regulations also apply under the *Food and Hygiene Act* where food and drink is supplied. It is prudent to seek the advice of the local fire officer, especially if competitions are being run, as this may come under the *Fire Precautions Act*.

An inspector from the local authority may make random checks on these establishments to ensure that the yard is operating within the Act, and furthermore he is empowered to visit any stable to see if a horse is being hired out without a licence. Employing a manager who is disqualified, failing to treat a sick horse, supplying a horse with defective tack, providing a horse that is unfit to work and making a false statement are other offences under the Act. There must also be a register kept of all horses under four years of age on the premises. It may

be easier to keep a record of all horses. Any animal at full livery is not covered by the Act but those at part livery are.

The *Horse Breeding Act 1958* requires stallions to be licensed under the relevant breed societies.

The *Animals Act 1971* covers such things as straying, carelessness and negligence and as such is a matter of interpretation for each individual case. Each case will invariably create a situation that will call for a determination on how the law will apply because of the fringe areas which come under this Act, such as if a horse leans over a fence and eats the neighbour's apples it may be considered to have trespassed without actually straying. On the other hand if the horse goes through a fence which belongs to the neighbour he cannot complain if he does not maintain his fence in a good enough state to keep out livestock.

Transportation of animals is covered by the *General Order 1973* which states that horses must be properly cared for in suitable transport and further to satisfy a vet that they are fit to travel overseas. Horses to be imported must have a Ministry Licence under the *Equine Animals Importation Act 1973*. It is illegal to export live horses for meat as specified by the *Export of Horses and Ponies Minimum Value Order*.

As long ago as 1949 an Act was passed to illegalise docking, nicking or any mutilation of the tail *(Docking and Nicking Act 1949)*. More recently the *Farriers' Registration Act 1975* came into being which requires anyone shoeing a horse to be trained and subsequently registered. This eliminates the 'cowboys' in the trade who were not formally trained but provided a service to the public.

Health and safety

It is the responsibility of employers and employees as well as self-employed persons to ensure that all reasonable care is taken at work in the stable yard. Each has a duty to the public to behave in a safe and careful manner to guard against possible accidents which are

185

liable to occur where horses are concerned. The *Health and Safety at Work Act 1974* demands that people are not exposed to unnecessary risks in their place of work and that as much precaution as possible is taken where a situation of high risk or danger to health is anticipated.

Protective clothing should be supplied by the employer if the employee is to carry out certain jobs. This can include overalls, helmets, gloves, shoes, goggles and dust masks. One hazard which has more recently been brought to the attention of those working with horses is the risk of contracting Farmers' Lung. This is caused by dust particles being inhaled and damaging the lungs which subsequently affects respiration and renders the victim short of breath. In time the patient will become partially disabled and no longer able to work in a dusty environment such as a feed shed, hay or straw barn, or indeed in a stable itself. As a precaution dust masks should be worn in a confined area, especially in an enclosed space such as a loft or barn where there are dust spores from bedding and fodder.

Her Majesty's Inspectors (HMI) are authorised to make visits to work places where there are employees to check all aspects of health and safety. They will advise the employer or his representative of any precautions or repairs which must be carried out in order to comply with the law. There are a number of leaflets available to explain the legal requirements as well as stickers to warn against possible hazards and remind workers of procedures. They will also explain the accident procedure and may ask to see the accident book. The keeping of such a book is a legal requirement where there are employees and the record must show details of all accidents which take place, including how they occurred, who the victim or victims were, when and where the accidents happened and the names of any witnesses present etc.

It is the responsibility of the owner to ensure that all buildings are in safe repair and that equipment is in good working order. Broken hinges or fasteners, loose fork handles and any worn or broken equipment are a safety hazard and must be repaired promptly. If an item cannot be repaired at once then it should be removed from the yard and labelled NOT FOR USE. The safety aspect of all tack and saddlery equipment has already been mentioned but cannot be over-emphasised. Doors and windows, partitions, mangers and fixtures or fittings will need maintaining in a safe state and not be an unnecessary risk to horse or owner. Horses themselves are a high-risk occupation without the additional risks presented by negligence on the part of the owner or staff.

Drains should be kept clean and disinfected and rubbish should be disposed of regularly to distract vermin. Stale feedstuffs also pose a health risk and will need to be discarded daily. The muck heap may be used as a compost heap and must therefore be sited at a suitable distance from the yard because of the health risk caused by vermin and flies.

Any movement (i.e. leading) of horses should be carried out in a safe, controlled situation. The handler is responsible for ensuring that all precautions have been taken. Moving horses to and from a paddock should be organised so that one horse is not left alone in the field. Traffic in the stable yard should be kept to a minimum with a speed limit of 5–10 mph; vehicles should preferably be stationary when young horses are being moved about. Ideally an enclosure around the boxes will at least contain the horses should any one get loose. It is well worth the expense of post-and-railing the area. Gates should be kept closed at all times, unless someone is on hand to stand near the gateway. It is too late to shut the stable door once the horse has bolted.

A novice rider should not jump a horse without someone nearby in case of an accident, particularly out hacking. In addition to the aforementioned safety checks on the tack, stirrups will need to be large enough for the rider's foot to slip in and out of freely. Stirrups too narrow can trap the rider's foot and cause a frightful accident. The rider's footwear must be appropriate, i.e. no wellingtons or shoes with heels.

Buying and selling horses

These transactions are partly governed by two Acts, the *Sale of Goods Act 1893* and the *Implied Terms Act 1973* but these do not cover all eventualities. However, a contract of sale, either in writing or verbally, is enforceable and should cover the horse's identity and the agreed price. Both the vendor and purchaser creates a legal obligation to each other and both must be above the legal age, have the authority to carry out the transaction and be of sound mind. The agreement may include provisos such as 'on the condition that' but must not be 'subject to'. The legal maxim is *'caveat emptor'* (let the buyer beware) whether he is purchasing privately or at auction. A veterinary certificate will satisfy the owner that the horse has passed an inspection and is in the vet's opinion suitable for the work required of him. A warranty is a description of a horse but does not cover obvious visible defects or bad conformation. Vices such as crib-biting, wind-sucking and weaving are considered an unsoundness and must be mentioned.

Road safety

Before riding or leading a horse on a public highway the owner should understand the law as it applies whilst he is responsible for a horse. Ensure that saddlery is safe and you know your horse. Keep to the left-hand side of the road, and when riding and leading, the led horse must be on the left. You must be able to control the horse in traffic, keep off footpaths and pavements and go with the traffic in a one-way street. When mounted you should wear a hard hat and after sunset carry lights which show red to the rear and white to the front. Light-coloured or reflective clothing, of which there is a choice available, is recommended to illuminate the rider after dusk. If the horse is traffic-shy it should first be taken out in the company of traffic-proof horses on quiet roads. Avoid peak periods and main roads. A hazard on the roadside should be given a wide berth after checking for traffic both ways and waiting until it is quiet.

The *Highway Code (1982)* describes more fully the law of the road and how it applies to uses with horses. It must be remembered that there are laws to abide by and the horseman is as responsible as any other road user. It is foolish for any rider to think that because he is mounted on a horse he is above the law, especially in the event of an accident. Children should be taught at an early age how to behave on the roads with their ponies, not forgetting simple courtesy to other road users. Pedestrians should be passed wide as they may be afraid of horses. Motorists who slow down for horses should be acknowledged. A third-party liability insurance cover will cover you against damage to other people's property. The horse should be shod if he is to be ridden on the road other wise he will become foot sore.

When riding in a group it is safer to have a traffic-proof horse in the lead and at the rear, each with an experienced rider, with any novice horses and riders in between in single file. The lead and tail rider should indicate simultaneously when they intend to cross the road or pull out to pass a parked vehicle. First they must ensure that the ride is not strung out and in danger of becoming separated. Never ride more than two abreast at any time and keep in single file on busy or narrow roads, approaching a bend or brow of a hill. Never carry out any manoeuvre until it is safe to do so keeping the horse as calm and quiet as possible at all times. The exception to this is with a young or nervous horse who should be ridden on the inside of one who is traffic-proof. This will encourage drivers to pass wider and slower which will give confidence to the traffic-shy horse. At all times the horse must be ridden straight, on a rein contact, with a whip in the rider's right hand. The horse should preferably have road studs in his hind shoes but in any event do not trot fast on main roads or those with a slippery surface. Remember once a horse is frightened by traffic he may never overcome his anxiety, so it is worth taking extra care and time to build up his confidence in traffic when he is young.

Riding and leading a horse at the same time is not advisable on major roads. If, however, it is unavoidable it should be carried out by an experienced rider with traffic-proof horses. The led horse should be fitted with a bridle, with the near-side rein passed through the bit and the animal led close to the lead horse's side, not getting in front or falling behind. With the rein attached in this way the bit cannot turn in the horse's mouth. A coupling ring attached to the bit on the end of a lead rope is an alternative to leading with a bridle. Always ride with the led horse away from the traffic, i.e. on your left-hand side.

Accident procedure

Accidents can happen to horses or people in the most carefully monitored situations, due to unforeseen circumstances. It is therefore necessary to adopt a safe practice in all that you do in the stable yard and indeed wherever horses are involved. First of all, don't panic in an emergency; learn to be calm and sensible; know the procedure to adopt and organise everyone efficiently.

The yard's first-aid kit must be accessible, kept clean and well stocked. The recommended minimum equipment is:

1. An assortment of about twelve adhesive wound dressings.
2. Sterilised unmedicated wound dressings in various sizes.
3. A triangular bandage for use as a sling.
4. A packet of cotton wool.
5. A sterilised eye-pad.
6. A few safety pins.
7. Adhesive plaster.

It is useful to have at least one person in the yard who is trained in first aid, particularly in a riding school. The more familiar everyone is with simple first-aid procedure the better, and new staff should be briefed accordingly. The telephone number of the local doctor, hospital, ambulance service and veterinary surgeon must be near to the telephone to enable anyone to dial in an emergency.

If a rider falls someone should attend to him/her while others concentrate on catching the horse. If the rider is sure that he feels all right and he has not suffered any form of head injury he may be allowed to remount in his own time and given the chance to recover fully. If, by chance, he may have hit his head he must not be allowed back on the horse until he has been attended to by a doctor, even if he was wearing a hard hat/skull cap. There is always a danger of concussion, which may be delayed. The old-fashioned principle of getting back on the horse as soon as possible may benefit the rider's confidence but can prove dangerous if he is injured. Minor wounds and abrasions will need attention as soon as possible and it should be established whether or not the injured person has a current anti-tetanus vaccination. Bleeding will need arresting with a handkerchief or clean piece of material folded into a pad.

If the rider is unable to move after a fall he must be kept warm and laid on his side to prevent him swallowing vomit. If conscious the site of pain can be established in order to convey as much information as possible to the doctor. Do not attempt to move the patient before the doctor arrives as this can often do more damage, particularly if a limb has been fractured or ribs broken. Apart from the patient's well-being there may be cause for an insurance or negligence claim, so it is essential that the proper procedures are followed and the patient protected against further injury. No matter what the degree of injury the patient will need reassurance so the accident procedure should be carried out in a calm and efficient manner.

A horse that has fallen or had a collision must be trotted up to ensure that he is sound before being re-mounted. If there are any minor wounds they should be bathed and treated accordingly. If they are serious the vet must be called in case there is a need for stitches. No treatment should be carried out until the vet has seen the wound but it can be cleaned with tepid salted water. As with a human, the horse's anti-tetanus injections will need to be kept up to date. If the horse is in open country or on the road and cannot be caught easily the police should be notified and help sought to catch the horse, which will likely head for home if he knows the area. If he has galloped and is blowing a capable person should lead him round quietly until his respiration has returned to normal, although this may not necessarily be convenient on the road side. If the horse is injured in a road accident, ask a passer-by to telephone for a vet and the police.

Mounting and dismounting

If a mounting block is available all horses should be taught to approach alongside and stand still parallel to it while the rider climbs onto the block and mounts the horse. The horse must be made to stand absolutely still until the rider asks him to walk on. If necessary have someone stand at the horse's head to help teach him. This lesson should be taught as early in the horse's education as possible and is the foundation for similar exercises such as mounting from the ground, standing in the show ring or opening a gate. The height of the

mounting block can be about 2–3 feet high with two or three steps. Ideally it should be made of stone and be a permanent fixture in the stable yard. Alternatively a solid wooden block is suitable providing it can be made stable. If it is light enough it may be useful to move it.

It is possible to have an extension on the stirrup leather which can be let down for the purpose of mounting and hooked back to the normal riding length once the rider is in the saddle. If the rider is mounting from the ground it is imperative, and this cannot be over-emphasised, that the horse from an early age be taught to stand square and still with his weight evenly distributed. He should not be allowed to move until the rider asks. Again, with a young horse an assistant may be needed to stand at the horse's head. It may require some time and patience at first but it is an important exercise which instils manners that will save the rider a lot of time and embarrassment in the future. Once mounted the rider

How to mount correctly with someone holding the horse.

should take his time to adjust the girth or stirrups if necessary. This will teach the horse to be patient and not anticipate moving off once the rider is aboard.

The horse is by nature wary of man standing on higher ground so that he has to look up and backwards at the person. However, careful training at an early stage will build up the horse's confidence and dispel any aversion to this. Whichever way you mount it is found that once the horse is used to being mounted from the block he will be easier to mount in open country. It also avoids the stirrup and foot touching the horse's belly when mounting while the rider searches for the stirrup.

A leg-up is when someone on the ground holds the rider's bent left leg with his right hand while the rider is facing the horse's near side opposite the saddle and lifts the leg as the rider jumps up and is raised into the saddle. The higher the leg can be lifted the easier it is for the rider to spring up and land gently in the saddle. Racing lads being lightweight are easy to leg up into the saddle. The rider should avoid sitting down heavily into the saddle especially if the horse is cold-backed.

To mount from the ground it is helpful to have someone hold the horse's head if he is young or nervous. Facing the tail on the near side the rider should hold the reins firmly in his left hand at the withers, place his left foot in the stirrup taking care not to touch the horse's side with his toe. Then levering with the left foot spring up and turn 180 degrees into the saddle with the help of the right hand on the back of the saddle. Again, the rider must take care to sit gently down into the saddle and put his feet into the saddle before riding forward. If the horse is objecting and tries to reverse or rein back the person on the ground should not hang on to the horse's mouth. This will only pull the bit forward and startle the horse even more.

It is equally important to take care when

How to give a rider a 'leg-up'.

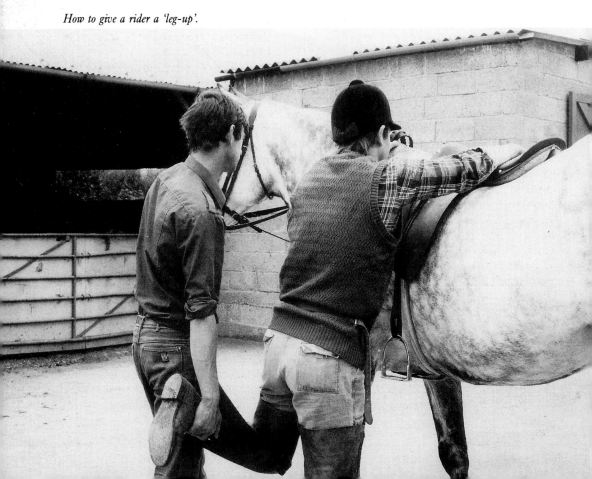

How to jump on a horse and lean over his shoulder before swinging the right leg over and sitting in the saddle.

dismounting. Even the quietest horse may suddenly take fright and shoot off. Young horses can pick up bad habits if not taught proper manners. They may jump away from the rider as he leaves the saddle which can be dangerous. At a standstill the rider should take a contact on the reins and hold them in his left hand on the withers. Taking his feet out of the stirrups he can then lean forward on the horse's neck and swing his right leg up clear of the horse's quarters, at the same time swivel his body round so that he is lying over the saddle before sliding gently down to the ground. If it is helpful to the novice rider he may hold a piece of mane in his left hand with the reins to help his balance. It may be useful with a young horse to have someone hold his head while the rider dismounts.

To hold a horse for someone to mount or dismount the assistant should stand in front of and face the horse, taking a rein in each hand, having contact with the horse's mouth. This way he should have sufficient control to stop the horse walking forward and discourage him from swinging his quarters sideways.

Mounting on the move is only suitable for gymkhana and Pony Club ponies who have to learn to do this in competition because it is difficult to make a horse stand still for mounting once he has learnt to do this at either the trot or canter. Experienced riders will be able to leap on while the horse or pony is going almost at a gallop. Racing lads particularly seem to have the knack of jumping into the saddle while the horse is either at a standstill or walking. It doesn't matter how tall the horse is if one has the spring in one's feet. Some people on the other hand, no matter how fit they are, have difficulty jumping on a pony because one has to have the knack of getting the left elbow well over the horse's withers which will enable the rider to throw his weight to the off side by leaning over and swinging the right leg round. It is a useful thing to practice when one is young but it becomes more difficult with age!

191

Loading a horse for transportation

It is an important part of the horse's education that he learns to go in a horse box or trailer calmly and confidently from an early age. A bad experience or fright connected with this can unsettle him permanently, making him a difficult loader and nervous traveller. It is not uncommon for a horse to be happy about loading and travelling in a horse box but unwilling to go into a trailer. This is often because the roof of a trailer is low and the horse feels he does not have enough room. Trailers are also noisy which is unsettling for the horse. Most horses travel better with a partition, even if they are alone. It is, however, illegal to travel more than one horse in a trailer without a partition.

Young horses should always have an experienced horse person to organise their loading because no one can be sure how they will react. The vehicle can be parked alongside a wall or gate so that the horse can only swing out to one side. All possible safeguards will be necessary to ensure that there are no hazards likely to distract a horse, such as other horses moving about, nearby traffic or any other activity which may unsettle him. Ideally the ramp should not be too steep or slippery. A sprinkling of straw or shavings on the ramp will help the horse get a grip. Some trailer ramps are fitted with coconut matting. Partitions must be secured to one side to allow easy access for the horse and handler. If possible the company of another horse that is a good loader and traveller will be helpful in encouraging the young or difficult horse to load and give him confidence to settle once he is on board. Do not leave the introduction to travelling to the day when a horse *has* to be transported. He will need a few lessons beforehand, when there is time and a calm atmosphere to develop confidence; not a hurried and tense atmosphere which is often inevitable on the day of a competition. Preferably, once the horse has been in and out several times and is happy about going into the vehicle he should be taken on a short journey very slowly. This will be an investment for the future so it well worth taking time and patience. If he is suddenly thrown into a situation where he has to make a journey in a hurry he may always be reluctant to load in the future.

A nervous or difficult horse will be easier to control if he is wearing a bridle (perhaps on top of a headcollar) with the reins not passed over his head. A lunge line can be used instead of reins, but never use a chain. The person leading the horse should hold the reins in one hand and a bowl of food in the other to encourage him to go quietly up the ramp. He must never look back at the horse when leading him, but instead encourage him to lower his head. Someone should be behind the horse to keep him going forward if possible. If necessary have a third person to lift the horse's front leg and place it onto the ramp as he stands quietly at the foot of it. Encourage with the voice and reward with a handful of food and a haynet. A purpose-built loading ramp is ideal but as far as possible do not load a horse in open countryside.

A particularly difficult loader may be encouraged by the use of a lunge line held at each end and passed around the horse's quarters as he stands at the ramp, being tightened as the horse moves forward. One must take care to keep the lunge line high enough so that the horse cannot become tangled in it if he kicks out. If a horse will not load easily at home doubtless he will not be any better in unfamiliar surroundings. Normally three strong handlers are enough to load most horses.

It is important to keep the horse as relaxed as possible throughout, once he becomes agitated problems begin and he may lose his temper and possibly kick out or rear up. There is, however, often a time when a stroke with a lunge whip around the horse's quarters will be sufficient encouragement for a stubborn horse to load. An experienced handler will soon establish whether a horse is genuinely nervous and needs to develop confidence or is just plain stubborn and awkward. In the first instance it will do more harm than good to frighten him with the use of a whip.

Assistants should be briefed as to how to handle the horse and not lose their tempers either if a horse is being difficult. Sometimes a standing martingale fitted on a horse who is prone to throwing up his head will give the handler better control. Once the horse is loaded be sure not to tie him short until the ramp is closed in case he runs back and ensure that no one is standing immediately behind the ramp in case the horse kicks or pushes the ramp back. Trailers without a front ramp are often the cause of a horse running back. It is better if the horse can see through the front and when he is first taught can be led straight through. It is advisable to attach cross ties to a horse in a trailer which will prevent him from turning round.

When unloading a young horse take him very quietly to the ramp and beware that he does not try to jump off from the top which can result in the horse falling on landing. Wherever possible choose a non-slippery surface on which to unload and discourage the horse from rushing. The slower this is done the safer it will be.

Insurance

It is a legal requirement that employers are insured against liability for personal injury or disease sustained or caused by their employees arising out of or in the course of their employment. This is known as Employers' Liability Insurance. The insurance certificate should be displayed at the workplace for employees to see. If horses are hired out to the public the proprietor of the riding establishment is obliged to be insured against public liability.

Horses, too, can be insured for almost anything from veterinary fees to loss of use. Policies vary, of course, but some include saddlery, the horse transport and its contents, i.e. any tack and equipment that may be stolen. Horse insurance can be costly depending on what the owner wishes to insure against if the horse is to compete. It is often well worth it for a few horses but it does become prohibitively expensive for a number. As with all insurance cover it is prudent to shop around until the best rates are found for the cover required. Rates are usually based on the value of the horse and the purpose for which it is to be insured. It may therefore be necessary to re-evaluate the horse annually in case an adjustment on the policy is required.

CHAPTER EIGHTEEN

STABLE VICES AND REMEDIES

Since the horse has been domesticated and stabled for long periods of time he has developed bad habits to alleviate his boredom. Some of these habits, such as weaving, appear now to have become hereditary. It is sometimes possible to overcome a habit at the outset but, by and large, it is virtually impossible to effect a permanent cure once it is established. The short answer is, of course, to avoid the horse becoming bored whilst he is stabled but this is not always practical. Each animal's temperament has a lot to do with the way he behaves in captivity. The stable routine will also affect him. Good management plays an important part in the horse's welfare and behaviour by limiting the stress that is placed on the horse at all times thereby reducing anxiety. This chapter deals with the most common vices and the methods of dealing with them.

Weaving

A horse is said to be a weaver when he sways his head sideways, to and fro, with or without raising his forelegs alternately off the ground in a monotonous rhythm. There are varying degrees of habituation which range from the animal who will only weave when he becomes anxious and excited, perhaps at feed time, and leans over the stable door, to the serious offender who, despite attempts to restrain him, will mindlessly weave for long periods at a time whilst he is confined to his stable. Once the horse is an established weaver it is safe to

say that he will not be cured and he will stand in the box weaving for no apparent reason, even without his head over the door. On the other hand the horse who only weaves for brief moments when he is stressed or nervous is often distracted successfully by placing an anti-weave bar over the stable door or suspending a heavy object, such as a brick, so that the horse will hit his head if he weaves over the door.

It is a vice brought about by nervous reactions and therefore usually found in the highly strung animal. Apart from being highly contagious and consequently something which young horses will quickly learn, it is a vice which must be declared at the time of sale or for veterinary purposes. Whatever the degree of seriousness of this habit it is unacceptable to most owners especially if the horse is unproven.

Crib-biting

This is another incurable habit, which is also harmful to the animal's health. Some crib-biters lose condition whilst others are known to gain weight. The crib-biter will grasp at a fixture, usually the top of the stable door or manger, with his teeth and suck in air. As he does so his nostrils dilate and the sound of the air intake is audible. Eventually the front incisors will show and may not meet properly, which will affect the horse when eating especially grazing. As a result his condition may deteriorate.

Even when all fittings are removed and a grille fitted above the door, the serious crib-biter will still find some way of committing this evil vice which is tolerated by very few people. One of the suggested causes of crib-biting is that the horse may have been under-fed when he was particularly hungry. Whilst it is an incurable vice once established there are deterrents available which are designed to inflict temporary pain to the horse when he tries to inhale. One of the most popular devices is a collar which is fitted tightly around the gullet to restrict the passage of air when the horse takes a deeper breath than normal at rest. Another is a similar strap fitted with pointed studs on the inside which press into the horse's throat when he inhales. Like most vices it is highly contagious amongst stable companions especially youngsters and aged horses. Boredom can also be a cause but, whatever the reason, once the horse is seen to crib-bite it must be declared as it is an un-soundness.

Wind-sucking

This is not unrelated to the above vice insofar as the animal sucks in air and swallows in the manner of gulping. The main difference is that he does not grasp hold of anything with his teeth to do so. The horse's condition will suffer eventually and, like crib-biting, it is an unsoundness which is quickly picked up by other horses. The only preventions are to remove all fittings from the box to discourage him from progressing to crib-biting and to fit a muzzle whenever the horse is stabled and not eating.

Rug and bandage chewing

Rug chewing is a habit, as distinct from a vice, and is infuriating from the owner's point of view in terms of the damage it can cause if the horse does not grow out of it. Young horses are the main offenders and will regularly pull their rug forward from under the roller until it is hanging to the ground so that they can then chew it or tread on it until it is torn and serves no purpose. A horse with a sensitive skin or one that becomes heated and starts sweating can find that a rug will irritate and try to remove it, especially when he lies down and is uncomfortable.

Apart from applying creosote or Cribox to the rug edges the alternative is to fit a cradle or bib. A cradle is made of a number of round pieces of wood roughly the length of a horse's neck. These are joined together with thin rope, hung around the animal's neck and tied to form a tubular shape. A leather or plastic bib, which is attached to the headcollar and hangs down about 12–18 ins (30–45 cm) will prevent most horses from reaching behind to grasp the rug. The horse is still able to feed and drink whilst wearing a bib and it is less restricting than a cradle. However, a cradle will absolutely prevent a horse from turning his head and reaching his rugs.

Bandage chewing can also be avoided by smearing on an offensive-smelling application such as Cribox or creosote. One word of warning here: whenever these substances are used be sure that the horse is not allergic to them because he may show a skin reaction on his muzzle which could be passed to other parts of the body. Boredom is the most common cause of bandage chewing so if the horse has more to think about in terms of his work and preventative measures are taken, the chances of him overcoming the habit become more likely. It has been suggested that digestive abnormalities may also be a cause of rug or bandage chewing.

It is impossible to wash creosote or Cribox off once it has been applied to bandages so it is cheaper to strap surgical tape once round the top and bottom of an old bandage and put the Cribox or creosote on that instead. The cradle is designed to prohibit a horse from touching his front legs with his muzzle and for this reason is often used on horses who are being given veterinary treatment such as firing, blistering or to stop them interfering with bandages.

Biting and kicking

Apart from the inconvenience to the handler these habits are inexcusable because as often as not they could have been overcome with correct handling during the horse's earliest education. Horses who are deliberately vicious in this way are becoming fewer nowadays, probably because youngstock are handled more consistently and proficiently than they used to be. There is, however, no room for complacency whenever horses are handled, no matter how familiar they are to you, and vice versa because even the quietest animal who has never been seen to kick or bite in ordinary circumstances can react in this way in times of stress or when suddenly frightened. The golden rule of always being alert when handling horses should never be disregarded because when a horse is startled he is likely to behave uncharacteristically.

If a young horse kicks or bites or even threatens to do so he should be reprimanded promptly and harshly because once he is allowed to get away with it he will do it again and in no time at all it will become a habit with him. Never strike a horse on the head but hit him anywhere behind the shoulder on his body, preferably under his belly. If necessary use a whip to punish him with a sharp stripe as soon as he kicks or bites. At the same time raise your voice to scold him for his misdeed so that he relates it to his behaviour.

Wall kicking and door banging

There is a risk of injury to the horse that kicks or strikes at the wall or door as well as the chance of him spreading a hind shoe. In addition there is a constant danger of him breaking wooden boards or doors. Apart from lining the walls and door with rubber matting to guard against injury and reducing the noise there is little one can do to cure him of the habit. Some horses only ever do it at feed times, out of impatience, whilst others may do it at any time just for something to do. Always take care not to threaten to hit a horse who is leaning over the door to bang it because he will raise his head in fright and possibly hit himself on the top edge. If the horse is sensible enough to be left with the door open and a chain fixed across the doorway instead this will at least save him from banging his knees although it won't stop him from pawing the ground.

GLOSSARY OF EQUESTRIAN TERMS AND EXPRESSIONS

AGED – a horse above seven years of age.

AZOTURIA (or 'Monday morning disease') – a condition where the horse's legs swell up, caused by being rested while still on a working diet.

BACKING – to back a horse is to teach him to accept a rider on his back.

BALANCE – the first requirement of schooling a horse. Horses may or may not have good natural balance which will affect the way in which they carry a rider.

BANGING OR STRAPPING – although the latter word refers also to grooming they both describe the act of stimulating the horse's muscles by regular and rhythmical slapping of the groom's hand in a particular area. It is carried out after the other duties of grooming have been performed to activate and develop muscles in the shoulder, neck and quarters of the horse, taking anything from up to half an hour on either side of the animal. In order that the horse should benefit fully from banging, the groom should take a hay wisp (see Chapter 9 for details of how to make one), a leather banger which is specially designed for the purpose or a stable rubber folded up into a wad. The animal should be made to stand square and still during this time. Any persistent resistance may be an indication of injury and banging must be stopped until the horse has recovered. The animal should, generally speaking, have reached the stage in his fitness programme where he is trotting before banging is incorporated into the grooming as his muscles may not be ready to withstand the repeated massage.

BLOOD HORSE – otherwise known as a Thoroughbred.

BOX WALKING – although not categorized as a vice but rather a bad habit, this is sometimes overcome once the animal is in work and has a schedule to occupy him. This habit takes the form of relentless walking around the box, making a pathway so that the animal is often left with areas without bedding.

BROKEN – a horse is said to be broken when he has been mouthed, backed and ridden away.

BROKEN KNEES – those which have been scarred by injury.

BROODMARE – a mare used for breeding.

BRUSHING – although this is a common fault it is often overcome as the horse develops and becomes fitter. Again conformation and weakness are the main causes. Brushing can occur both in the fore and hind limbs mainly around the fetlock joint, when one fetlock or hoof chafes against the other as the horse moves. Leg protectors of some kind are essential whilst the horse is being exercised.

BY – sired by.

CLENCH – a farriery term which describes the nail tip that has been driven through the shoe and the horse's hoof and is then bent over and fastened down with a nail clencher or hammer. With wear these clenches rise and can cause injury because they are sharp. They are easily hammered down again as a temporary measure whilst waiting for the farrier to deal with the shoe.

CLOSE TO THE GROUND – describes a short-legged horse with a deep body.

CRADLE – a wooden frame fitted to the horse's

neck to prevent him turning his neck and reaching his legs. Often used after veterinary treatment such as blistering or firing.

CRIB-BITING – an unsoundness and usually an incurable habit where the animal grasps hold of a fixed object, usually a stable door or fitting such as a manger, with his teeth and holds onto it. There are devices on the market to counteract it but it is rare for a serious offender to tire of this vice.

CROP – a riding whip mainly used for hunting.

CRUPPER – a piece of tack fitted from the horse's dock to a roller or the back of a saddle to prevent the roller or saddle slipping forward. Useful on overweight ponies with badly fitting saddles.

CUB-HUNTING – begins at the end of August when young hounds are taught to catch young foxes.

CURBY HOCKS – hocks with poor conformation which may predispose to curbs.

DAISY CUTTER – refers to a horse with low action which causes the toe to flick along the grass.

DAY RUG – a woollen rug used either in the stable during the day or as a travelling rug.

DE-NERVED – the nerve supply to the foot is sometimes severed when a horse is suffering from chronic foot lameness. This method can be used to relieve pain, as in navicular disease. A de-nerved horse is unsound.

DISH FACED – a concave-shaped face usually found in Arabs.

DISHING – faulty action when the horse's fore leg/s are thrown outwards at any pace.

DOING-UP – the time for doing-up or evening stables is usually between 4–5 pm. The duties include skepping out, tidying the yard, renewing water, giving hay, changing rugs, changing bandages and administering medicines.

DRAUGHT HORSE – a heavy breed of horse common in Ireland.

DRENCH – a form of administering drugs orally.

ENTIRE – another name given to a stallion.

FLASHES – a decoration on the horse's quarters made with a body or water brush.

FORE – the front part of the horse. Fore legs are front legs.

FORGING OR CLICKING – weak horses and those who are allowed to slop along off the bridle are the main culprits at this but they often overcome it as they become fit and or are ridden properly. The toe of the hind hoof strikes the shoe of the front hoof and makes a clicking noise. It can eventually cause a front shoe to become loose if allowed to continue.

FRESH – usually refers to a high-spirited horse.

FULL BROTHER/SISTER – progeny having the same sire and dam.

FULL MOUTH – a horse with all his permanent teeth.

GENERAL STUD BOOK – registration book for Thoroughbred horses.

GOING SHORT – the action of a horse who appears to be taking shorter strides with his front legs, indicating some discomfort which can lead to lameness.

GREEN – an inexperienced young horse who, although broken, is not fully trained.

GRIPES – an old-fashioned word for colic.

GROWN OUT – in showing this means a horse or pony has exceeded the height limit for a class.

HALF-BRED – a horse of mixed, common breeding.

HALF BROTHER/SISTER – progeny with either the same sire or dam.

HALTER – usually made of webbing and an alternative to a headcollar. The nose band forms a running loop with the rope or chain to which it is attached.

HEAD LAD – another name for the head groom in a racing stable.

HEART ROOM – the space through the horse's girth and chest; a horse has good heart room if his girth is deep and his chest is broad and open.

HEEL-BUG – caused by wet muddy conditions usually affecting horses with white heels. Often carries an infection that can set up an allergy and cause lameness as the heel's fetlock joint swells up.

HERRING-GUTTED – a horse with a weedy narrow body which runs up sharply from the girth to the hindquarters. These horses always look worse after strenuous exercise.

HIGH-BLOWING – caused by a flapping of the

false nostril which is more easily heard when the horse is working at faster paces. It is often considered by the novice horseman to be an unsoundness of the horse's wind.

HIND – the back part of the horse. Hindquarters are the horse's bottom.

HIRELING – a term used for horses that are hired out, particularly for hunting.

HOBDAYED – describes a horse with respiratory problems whose larynx has been operated on. The name comes from the inventor Sir Frederick Hobday. It is a requirement by law that the owner declares that his horse has had this operation if he is offered for sale.

HORSE SICK – pastures which have become soured by grazing for too long and are suspected of having a heavy infestation of parasites.

HOSTLER – an old-fashioned name for a groom.

HUNTING GATE – a narrow wicket gate wide enough for one rider.

HUMOUR – a skin condition caused by the horse's blood over-heating.

LASH – the silk or cord which is attached to the thong of a hunting or driving whip.

LAWN MEET – a meet of hounds at a house by invitation of the owner.

LET DOWN – refers to the horse being taken out of work or training to rest.

LOINS – part of the horse's back, behind the saddle.

LONG REINS – lunge lines or similar, about 20–30 ft (6–9 m) long, suitable for long-reining a horse.

LONG IN THE TOOTH – refers to an old horse.

LOOSE BOX – is a stable, so called to differentiate it from stalls wherein the animals were tied continually.

LOP EARS – these hang downwards to the side giving a droopy appearance. Often found in Thoroughbreds.

MADE – an expression given to an educated or schooled horse indicating that he is ready for an average rider to ride, e.g. a made hunter.

MAIDEN MARE – a mare who is yet to foal for the first time.

MAKES A NOISE – means that a horse has a respiratory problem which is audible.

MANÈGE – a dedicated schooling area.

MFH – Master of Foxhounds.

MOUTHING – the process of bitting a horse to control him with a bit in his mouth.

NAILBIND – this occurs when a farrier has nailed too close to the sensitive part of the foot causing lameness.

NEAR – the left side of the horse.

OFF – the right side of the horse.

ONE SIDED – describes a horse that works better on one rein than the other perhaps as a result of muscle stiffness or a problem in his mouth.

OVER-REACHING – a horse is said to have over-reached when he has struck a fore leg with the toe of a hind hoof. Conformation, weakness or over-tracking when jumping are the main causes. Over-reach boots will help protect the heel and coronet area and tendon protectors, brushing boots or bandages will shield the leg between the knee and fetlock.

PARROT MOUTH – a deformity of the upper jaw when the top front teeth overhang the lower jaw preventing an even contact. In extreme cases it prevents the horse from grazing easily.

PEACOCKY – an unnaturally high head carriage which sometimes appeals to the novice horseman.

PIGSKIN – a type of leather sometimes used for the seat of quality saddles.

PIPE OPENER – referred to when a horse is given a sharp gallop to clear his wind-pipe.

PLAITING – describes the movement of a horse that places each front foot in front of the other and does not maintain a straight action from the shoulder through the knee.

PLATE – a light shoe usually made of aluminium and used for racing and showing.

POLL EVIL – a painful soft swelling on the horse's poll caused by a blow to the head.

PRESENCE – a term used to describe the horse's disposition. A lovely presence is sought in showing horses and ponies because it is eye-catching.

QUARTER MARKS – a cosmetic turn-out feature of check patterns on the horse's quarters, made with a water brush and fine comb.

RACING PLATE – an aluminium shoe used for racing and sometimes for showing. The horse is usually reshod with his normal shoes after a race.

RACKING UP – a term adopted for tying up a horse, which harks back to the use of a rack chain for this purpose. More often today a lead rope is used. Historically the act of racking-up was a main feature of stable management and horses were not attended to before this was done. Proper rack chains can still be bought but care must be taken when using them to see that the horse does not catch his teeth when playing with them. They should not be left in the stable for the same reason.

RANGY – a term used to describe a tall horse with lots of scope.

RAT CATCHER – riding dress comprising a tweed coat and dark breeches with a bowler hat and long boots, used for showing and cub hunting.

RIDDEN AWAY – the stage at which a young horse, freshly backed, is now able to hack out alone.

RIG – a gelding which has been castrated but has retained one testicle. Rigs are often a nuisance because they still look for mares and often have an inconsistent temperament.

ROUGH OFF – a transitional process from a state of fitness to one of the horse being able to live out while he is on holiday. Otherwise referred to as letting down.

SCARLET – the correct term for a red hunting coat.

SCOPE – describes a horse's potential and is recognised by experienced horsemen.

SEASON – another word for oestrus.

SEATED-OUT SHOE – when the flat surface is hollowed out so that it does not make contact and put pressure on a particular part of the horse's sole.

SERVICE – the mating of a mare with a stallion.

SET FAIR – to set fair a bed once it has been mucked out is to rearrange the bed and renew the old straw with fresh before laying it out evenly with banks all around the walls. Banks are a safeguard against the horse rolling and becoming cast or catching his hocks. The bed should be set fair before the horse returns from exercise and at 'doing-up' time.

SHELLY FEET – brittle, small and thin-soled feet.

SHORT-COUPLED – a horse which is short in his back, deep in the body with well-sprung ribs. Often descriptive of a good sort of horse.

SICKLE HOCKS – bent, weak hocks which do not stand up to hard work.

SINGEING – a means of burning cat hairs. Singeing lamps are becoming difficult to find because they have been superseded by clippers. A candle is suitable for the job but requires experience to do a neat job.

SKEPPING OUT – this term refers to collecting droppings from the horse's bed with a skep (a receptacle designed for the purpose). To skep out properly the groom should not take out any clean bedding unnecessarily. The skep should be taken in with you whenever you attend a horse so that droppings are never accumulated in the box. It is also good practice to use the skep when picking out the horse's feet so that mud and stones do not gather in the box.

SLIP-STRAW – traditionally, once a stable or stall had been mucked out it was then covered with a thin layer of straw – known as slip-straw – to enable the floor to air whilst giving the horse some protection from slipping. It also encourages him to stale when he might otherwise not do so without any bedding underneath him.

SORE SHINS – a condition whereby the horse's cannon bones are sore as a result of concussion or injury.

SPECTACLES – another word for an Irish martingale.

STAR GAZER – a horse that holds his head very high which can be dangerous for jumping.

STUD GROOM – the head groom, usually in a stud.

SWALES GAG – a gag bit used to keep the horse's mouth open. Often used by veterinary surgeons.

SWAY BACKED – a horse that is hollow behind the withers, similar to dippy backed but often comfortable to ride.

THATCHING – this old expression refers to the method of drying off a horse which has returned from work still hot, whereby his loins are spread with clean, dry straw and covered either with a jute rug worn inside-out or a sweat sheet. He can then be left for a short while until he is dry and sweat is no longer evaporating. In the case of horses who have a tendency to break out into a cold sweat the straw may have to be renewed from time to time to ensure that it is dry enough to soak up the moisture.

TOM THUMB BIT – a short-cheeked Weymouth.

TONGUE GRID – a narrow metal plate with a cheek strap which fits over the horse's poll underneath a bridle to prevent the horse getting his tongue over the bit.

TUCKED-UP – a term used to describe when a horse's stomach is tucked-in, as when a human holds his breath.

TURNED OUT – can mean two things, either to put in the field or a horse's appearance, e.g. a well turned out horse is one who is well groomed and presented.

UNNERVED – a term used when part of the horse's nerve supply has been cut off for veterinary purposes.

UP TO WEIGHT – describes a horse that is classified as a weight-carrying animal, such as a hunter or show hunter, and truly qualifies as being able to carry that weight.

VICE – a bad habit which may or may not be regarded as an unsoundness.

WALL EYE – an eye with a white/blue colour.

WARRANTY – a guarantee given to a horse sold as sound in wind, eyes, heart and action and capable of doing the job for which he has been purchased.

WEAVING – is when the horse swings his shoulders, neck and head from side to side often relentlessly. More serious offenders raise alternate fore legs off the ground in a monotonous rhythm. There is no cure although anti-weave bars of various designs can restrain the animal to a degree. Some horses will stand within the stable to weave whilst others will only do it over the door.

WEED – an expression used to describe a small Thoroughbred horse with poor conformation.

WHITE LINE – a band of soft horn between the wall and sole of the foot. It indicates to the farrier the amount of wall he has in which to place the nails.

WHORL – a circular area of the horse's coat mentioned in descriptions for identification purposes.

WIND-SUCKING – this is a vice which stems from crib-biting whereby the animal will take hold with his teeth of a solid object and proceed to suck in wind. Not only is it an unsoundness but it is also harmful to the animal for as he sucks, his lungs expand and therefore his chest cavity and surrounding muscles. The physical effect is a loss of condition in the long run. There are a number of items available which can be fitted to the animal whilst he is stabled but once the habit is established it is almost impossible to cure permanently.

WOOD-CHEWING – horses will chew wood wherever they can find it, be it the stable door or a tree. Often a repellent applied to the wood will deter the animal from doing this. Owners should always be alert to the possibility that the horse which starts to chew or bite wood can develop into a crib-biter.

INDEX